Urban Sustainability

The Urban Sustainability Book Series is a valuable resource for sustainability and urban-related education and research. It offers an inter-disciplinary platform covering all four areas of practice, policy, education, research, and their nexus. The publications in this series are related to critical areas of sustainability, urban studies, planning, and urban geography.

This book series aims to put together cutting-edge research findings linked to the overarching field of urban sustainability. The scope and nature of the topic are broad and interdisciplinary and bring together various associated disciplines from sustainable development, environmental sciences, urbanism, etc. With many advanced research findings in the field, there is a need to put together various discussions and contributions on specific sustainability fields, covering a good range of topics on sustainable development, sustainable urbanism, and urban sustainability. Despite the broad range of issues, we note the importance of practical and policy-oriented directions, extending the literature and directions and pathways towards achieving urban sustainability.

The series will appeal to urbanists, geographers, planners, engineers, architects, governmental authorities, policymakers, researchers of all levels, and to all of those interested in a wide-ranging overview of urban sustainability and its associated fields. The series includes monographs and edited volumes, covering a range of topics under the urban sustainability topic, which can also be used for teaching materials.

Gaetan Siew · Zaheer Allam · Ali Cheshmehzangi
Editors

Diversity as Catalyst: Economic Growth and Urban Resilience in Global Cityscapes

Springer

Editors
Gaetan Siew
Port Louis, Mauritius

Ali Cheshmehzangi
School of Architecture, Design
and Planning
The University of Queensland
Brisbane, QLD, Australia

Zaheer Allam
School of Architecture, Design
and Planning
The University of Queensland
Brisbane, QLD, Australia

ISSN 2731-6483 ISSN 2731-6491 (electronic)
Urban Sustainability
ISBN 978-981-97-8775-3 ISBN 978-981-97-8776-0 (eBook)
https://doi.org/10.1007/978-981-97-8776-0

This Springer imprint is published by the registered company Springer Nature Singapore Pte Ltd.
The registered company address is: 152 Beach Road, #21-01/04 Gateway East, Singapore 189721, Singapore

For "urban activists" of all backgrounds and at all levels around the globe.

Preface

"We cannot act local without thinking global".
Address by Mr. Gaetan Siew, Sao Paulo, Brazil, November 2007

In the contemporary landscape where global cities stand as crucial hubs for both economic progress and cultural intermingling, the essence of this book lies in its forward-thinking approach to comprehending and utilizing the intricate montage of diversity found in urban settings. This exploration ventures beyond viewing cultural diversity merely as a social asset, positing it instead as a fundamental engine driving economic prosperity and urban resilience. Hence, the combination is what we propose and what we advocate for the future of cities as well as future cities.

This book distinguishes itself through its interdisciplinary vantage point, weaving together insights from urban planning, economics, social sciences, and sustainability studies. Such a holistic approach offers a novel perspective on the dynamics of global cities, enabling a deeper analysis and understanding that align seamlessly with Sustainable Development Goal 11, which aims to make cities inclusive, safe, resilient, and sustainable. The narrative of this book spans both breadth and depth, addressing an array of subjects from the economic advantages of cultural diversity in metropolitan areas to methodologies for harmonizing cultural heritage with contemporary urban development. This makes it an indispensable resource not only for academics and policymakers but also for urban planners and anyone keen on the evolution of urban living.

One of the key strengths of this book is its fusion of theoretical and practical elements. By providing actionable insights, it guides readers on how to effectively leverage cultural diversity in urban development. This blend ensures that readers are not only informed by theory but also equipped with practical strategies for real-world application. Through the inclusion of case studies and examples from cities across the globe, this book maintains a relevance that spans diverse geographic and cultural contexts. These real-world illustrations not only enrich the theoretical discussions but also demonstrate the practical implementation of the ideas presented, making the content accessible and applicable to a global audience.

Furthermore, the book's direct alignment with global policy agendas, especially the United Nations' Sustainable Development Goal 11, positions it as a strategic guide for policymakers and urban developers. We believe in the movement that suggests new pathways and paradigms for people-centric development and not mere development. By offering a roadmap to making cities more inclusive, safe, resilient, and sustainable, the book then highlights the fundamental role of cultural diversity in achieving these objectives.

As cities around the world continue to grow and evolve, understanding the multifaceted impacts of cultural diversity becomes increasingly crucial. This book highlights how diversity can act as a catalyst for innovation, economic growth, and societal resilience. By examining the interplay between cultural diversity and urban development, it reveals how cities can harness their diverse populations to create more vibrant, dynamic, and resilient urban environments. In a world where migration and globalization are constantly reshaping the urban fabric, the insights provided in this book are timely and pertinent. It addresses the challenges and opportunities that arise from cultural diversity, offering strategies for fostering inclusive communities and promoting sustainable urban development.

Ultimately, this book serves as a comprehensive guide to the benefits and potentials of cultural diversity in cities. It challenges readers to rethink traditional notions of urban development and to embrace diversity as a key driver of innovation and resilience. Whether you are an academic, a policymaker, an urban planner, or simply someone interested in the future of urban living, this book offers valuable perspectives and practical tools for navigating the complexities of urban diversity.

As we move forward in an increasingly interconnected world, the ability to understand and harness the power of cultural diversity will be essential for building cities that are not only economically prosperous but also socially inclusive and environmentally sustainable. This book provides the foundation for this understanding, equipping readers with the knowledge and strategies needed to create urban environments that thrive on diversity.

Lastly, we recognize the importance of diversity in cities, without which we can only think and act local and potentially ignore addressing global challenges and issues that we are the cause and part of. With this in mind, we advocate the significance of people-centric urban development now and in the future.

Brisbane, Australia Ali Cheshmehzangi
Brisbane, Australia Zaheer Allam
Port Louis, Mauritius Gaetan Siew

Acknowledgements We acknowledge everyone's contributions to this exciting book project. We extend our gratitude to all the authors, co-authors, chapter contributors, and collaborators who have shared their expertise, passion, and creativity with us.

This sentiment applies to our earlier work as well, where we deeply value the dedication and vision of all contributors that helped create this book. Such collaborative efforts foster new dialogues, networking opportunities, and a valuable research exchange platform for scholars worldwide.

About This Book

In an era where global cities emerge as pivotal hubs for economic growth and cultural fusion, this book presents an innovative approach to understanding and leveraging the rich tapestry of diversity in urban environments. This book delves into how cultural diversity acts not just as a social asset but as a key driver of economic prosperity and urban resilience.

What sets this book apart is its multidisciplinary perspective, combining insights from urban planning, economics, social sciences, and sustainability studies. It offers a fresh lens through which the dynamics of global cities are viewed, analyzed, and understood, aligning closely with Sustainable Development Goal 11—making cities inclusive, safe, resilient, and sustainable.

The scope of the book is both broad and deep, covering topics from the economic impact of cultural diversity in urban centers to strategies for balancing cultural heritage with modern urban development. It's a must-read for academics, policymakers, urban planners, and anyone interested in the future of urban living.

Key Features and Benefits

- Offers a unique blend of theory and practice, providing actionable insights for harnessing cultural diversity in urban development.
- Includes case studies and examples from cities around the world, making it relevant to a global audience.
- Directly aligns with global policy agendas, particularly SDG 11, offering a strategic guide for policymakers and urban developers.

Contents

About the Editors

Gaetan Siew is the Past President of the International Union of Architects and Past Secretary General of the African Union of Architects. Gaetan is an avid world traveller, experiencing over 500 cities across 105 countries including most of Africa. For his contributions to the Architectural and Urban world, he was awarded several honorary membership and fellowships, and elevated to the rank of Grand Officer of the Order of the Star and Key of the Indian Ocean by the Government of Mauritius, and elevated to the ranks of Chevalier de l'Ordre National du Mérite, and de l'Ordre des Arts et des Lettres by the French Republic. Past Board Director of Futures Cities UK since July 2013, Past Chairperson of State Land Development Company (Smart Mauritius), Past Chairman of Construction Industry Board, and current chairperson of the Port Louis Development Initiative and Special Envoy for UN Habitat, Mr. Siew advocates for sustainable approaches toward urbanism aimed to socially and economically regenerate urban fabrics.

Zaheer Allam holds a Ph.D. in Humanities, a Master of Arts (Res), an MBA, and a Bachelor of Applied Science in Architectural Science from universities in Australia and the United Kingdom. Based in Mauritius, he was the first Chairperson of the National Youth Environment Council (NYEC) and a board member of the Mauritius Renewable Energy Agency (MARENA) and works on a number of projects on the thematic of Smart Cities and on strategies dwelling in the increasing role of technology in Culture and the Society. He is an Adjunct Associate Professor at the School of Architecture, Design and Planning (ADP) at The University of Queensland, Brisbane, Australia. Zaheer is also the African Representative of the International Society of Biourbanism (ISB), member of the Advisory Circle of the International Federation of Landscape Architects (IFLA), and a member of a number of other international bodies. Honorary Fellow at Deakin University (Australia), he holds a number of awards and commendations, including an elevation, by the President of Mauritius, to the rank of Officer of the Order of the Star and Key of the Indian Ocean, the highest distinct order of merit in Mauritius. He is the author of over 175 peer-reviewed publications and author of 14 books on the subject of Smart, Sustainable and Future Cities.

Ali Cheshmehzangi is a Full Professor and Head of the School of Architecture, Design and Planning (ADP) at the University of Queensland, where he leads the School and researches across the fields of architecture, design, and planning/urbanism. He has been in the World's top 2 per cent field leader since 2022, recognized by Stanford University. He is among the top 30 global scholars in the urban sustainability research area. Ali has a wealth of experience, expertise, and a strong commitment to innovation and excellence in the fields of architecture, design, and planning, the three core disciplines of his School at UQ. With a career spanning over two decades, he has made significant contributions to the academic and professional communities, with a focus on sustainable and environmentally conscious design. So far, Ali has published over 500 journal papers, articles, conference papers, book chapters, and reports. He also has 29 other academic books, three of which have received awards at the national, provincial, and municipal levels. He also has received international awards and recognition for his research on urban resilience studies and sustainability research, as well as a Vice-Chancellor's award for his impactful contribution to higher education.

Chapter 1
Diversity in Cities and Addressing Urban Sustainability

Ali Cheshmehzangi, Zaheer Allam, and Gaetan Siew

Abstract This introductory chapter explores the multifaceted role of cultural diversity in shaping urban environments. As cities worldwide become central to economic growth and cultural integration, this book provides a comprehensive analysis of how diverse urban populations contribute to economic vitality and enhance resilience against various challenges. Distinguished by its interdisciplinary approach, the book synthesizes perspectives from urban planning, economics, social sciences, and sustainability studies. This holistic viewpoint reveals the complex interactions that define contemporary urban life and highlights the importance of inclusivity and resilience as outlined in Sustainable Development Goal 11 (or SDG 11). The book's extensive scope examines the economic benefits of cultural diversity, strategies for preserving cultural heritage amidst modern development, and the overall impact of diversity on urban sustainability. By weaving together theory and practical examples, it offers valuable insights for academics, policymakers, urban planners, and anyone invested in the sustainable development of cities. Key features of the book include its combination of theoretical frameworks with practical applications, providing readers with actionable strategies for leveraging cultural diversity in urban settings. The inclusion of global case studies enriches the discourse, making the book relevant to a worldwide audience. By aligning closely with global policy agendas, especially SDG 11, it serves as a crucial guide for those looking to navigate and shape the future of urban living in a sustainable and inclusive manner.

A. Cheshmehzangi (✉)
School of Architecture, Design and Planning (ADP), The University of Queensland, Brisbane, QLD, Australia
e-mail: A.Chesh@uq.edu.au

Z. Allam
Urban Transformations Research Centre (UTRC), Western Sydney University, Parramatta, NSW, Australia

School of Architecture, Design and Planning (ADP), The University of Queensland, Brisbane, QLD, Australia

G. Siew
International Union of Architects (UIA), Paris, France

Global Creative Leadership Initiative (GCLI), Port Louis, Mauritius

G. Siew et al. (eds.), *Diversity as Catalyst: Economic Growth and Urban Resilience in Global Cityscapes*, Urban Sustainability, https://doi.org/10.1007/978-981-97-8776-0_1

Keywords Urban sustainability · Cultural diversity · Economic growth · Urban resilience · Sustainable development goals (SDG 11) · Urban planning · Inclusive cities

1.1 Introduction: Diversity in Cities

The modern world witnesses an unprecedented rise of global cities as central hubs of economic activity and cultural integration [28]. These urban centers not only drive economic growth but also serve as melting pots of diverse cultures, contributing to the rich social fabric that characterizes contemporary urban life [6, 9, 34]. This chapter provides a brief narrative and delves into the critical role that cultural diversity plays in shaping sustainable and resilient urban environments. This comprehensive exploration highlights the complex benefits of diversity and provides strategic insights for fostering inclusive, economically vibrant, and sustainable cities.

Urban sustainability is a pressing concern in today's rapidly urbanizing world. As cities expand and populations grow, the challenges of managing resources, ensuring environmental protection, and promoting social equity become increasingly complex [4]. This book posits that cultural diversity is a key asset in addressing these challenges. By integrating diverse perspectives and practices, cities can develop more innovative and effective solutions to sustainability issues [44]. The rich tapestry of cultural backgrounds within urban populations enhances creativity and problem-solving capabilities, leading to more resilient and adaptable urban systems [35, 38, 46]. Central to the discussion of urban sustainability is the concept of cultural diversity [5, 37]. Cities with a high degree of cultural diversity benefit from a broader range of ideas, traditions, and approaches to living and working [23]. This diversity fosters a dynamic and vibrant urban culture that attracts talent, stimulates economic activity, and enhances the quality of life for all residents [21]. Moreover, culturally diverse cities are better equipped to navigate and adapt to economic, social, and environmental changes [3]. By embracing and promoting diversity, urban centers can build more robust and sustainable communities.

One of the key aspects of concern related to diversity in cities is the significance of economic growth in cities [16]. Indeed, it is intricately linked to the presence of diverse populations. Cultural diversity drives economic development by attracting a wide range of skills and talents, fostering innovation, and creating new business opportunities [8, 19, 25]. Diverse urban populations contribute to a more dynamic and competitive economy, as they bring unique perspectives and experiences that can lead to the creation of new products, services, and industries [14, 45]. Additionally, culturally diverse cities are often more appealing to international businesses and investors, further boosting economic growth and development.

Urban resilience is another critical aspect of sustainable city development. Resilient cities can withstand and recover from various shocks and stresses, such as economic downturns, natural disasters, and social conflicts. Cultural diversity, or diversity in cities in general, enhances urban resilience by providing a wealth of social capital and community networks that can be mobilized in times of crisis. Diverse communities tend to have stronger social cohesion and support systems, which are essential for coping with and recovering from adverse events. By fostering inclusivity and social integration, cities can strengthen their resilience and ensure long-term sustainability. This statement correlates well with the concept of sustainable development, which is encapsulated in the United Nations' Sustainable Development Goals (SDGs), particularly SDG 11; i.e., a particular SDG that aims to make cities inclusive, safe, resilient, and sustainable. This book aligns closely with the principles and objectives of SDG 11 [10, 15], offering a strategic framework for policymakers, urban planners, and stakeholders to create more sustainable urban environments [17, 33]. By highlighting the importance of cultural diversity in achieving these goals, the book provides practical guidance on how to integrate diversity into urban planning and development processes.

We also note that urban planning plays a crucial role in shaping the future of cities [2]. Its role is even more complex in places or cities/communities where diversity levels are low. Effective urban planning must consider the diverse needs and preferences of urban populations to create inclusive and equitable urban spaces [39, 48]. This book emphasizes the importance of participatory planning processes that engage diverse communities and incorporate their perspectives into decision-making. By doing so, urban planners can ensure that development projects are more responsive to the needs of all residents, leading to more sustainable and inclusive outcomes [32]. In a way, urban planning fits the ideals of inclusive cities from multiple perspectives. Inclusive cities are those that embrace diversity and provide equal opportunities for all residents, regardless of their cultural background. Inclusivity is a foundation of urban sustainability, as it promotes social cohesion, reduces inequalities, and enhances the overall quality of life [22, 30]. This book explores various strategies for creating inclusive urban environments, such as promoting affordable housing, ensuring access to essential services, and fostering social integration. By prioritizing inclusivity, cities can harness the full potential of their diverse populations and create more vibrant and sustainable communities.

In sum, this chapter briefly offers a comprehensive analysis of the interplay between cultural diversity and urban sustainability. Through a multidisciplinary lens, the book highlights the economic, social, and environmental benefits of diversity and provides practical insights for leveraging this asset in urban development. By aligning with global policy agendas, particularly SDG 11, the book serves as a strategic guide for policymakers, urban planners, and stakeholders committed to creating inclusive and sustainable cities [11]. The future of urban living depends on our ability to embrace and harness the power of diversity, and this book provides the roadmap for achieving that vision.

1.2 Aim and Objectives of the Book

Our overarching aim is to provide a comprehensive understanding of how cultural diversity can be harnessed to promote economic growth, enhance urban resilience, and achieve sustainable development in global cities. Our objectives are stretched in a fourfold. The first is to explore the role of cultural diversity in economic development. In doing so, the contributions investigate how diverse cultural backgrounds contribute to economic vitality in urban centers, including the ways in which they foster innovation, attract talent, and stimulate new business opportunities [12]. The second is to analyze urban resilience in diverse cities. In doing so, contributing chapters examine how cultural diversity enhances the ability of cities to withstand and recover from various challenges, such as economic fluctuations, natural disasters, and social conflicts, by leveraging social capital and community networks. The third is to integrate multidisciplinary perspectives on urban sustainability. We believe this may be hidden in what may be highlighted but will help combine insights from urban planning, economics, social sciences, and sustainability studies to provide a holistic view of the dynamics of global cities, aligning with Sustainable Development Goal 11 to make cities inclusive, safe, resilient, and sustainable. And fourth, is to ensure the book offers practical strategies for inclusive urban development. Thus, our global contributions present actionable recommendations and case studies from around the world to guide policymakers, urban planners, and stakeholders in creating inclusive urban environments that balance diversity with urban sustainability, cultural heritage with modern development needs, and economic growth with urban resilience in global cityscapes.

1.3 Structure of the Book

The book is an accumulation of case study chapters from contributions around the globe. There are excellent examples of policy direction, practical pathways, and people-centric urban development patterns. Considering diversity as a catalyst in cities is crucial for fostering vibrant, innovative, and resilient urban environments [7, 18, 24]. Diversity encompasses various dimensions, including cultural, ethnic, socioeconomic, and professional differences (Fig. 1.1). These diverse perspectives can significantly enhance creativity, drive economic growth, and strengthen social cohesion in urban settings. For instance, cultural diversity can lead to a richer, more vibrant cultural scene, attracting tourists and boosting the local economy. Ethnic diversity often brings a variety of culinary traditions, festivals, and artistic expressions, enriching the cultural fabric of the city. Socioeconomic diversity ensures a mix of skills and talents, fostering a more dynamic and flexible workforce [31, 42, 47]. It also encourages more equitable economic development by creating opportunities across different segments of society [29]. Professional diversity, on the other hand, brings a range of expertise and innovative approaches to problem-solving, driving

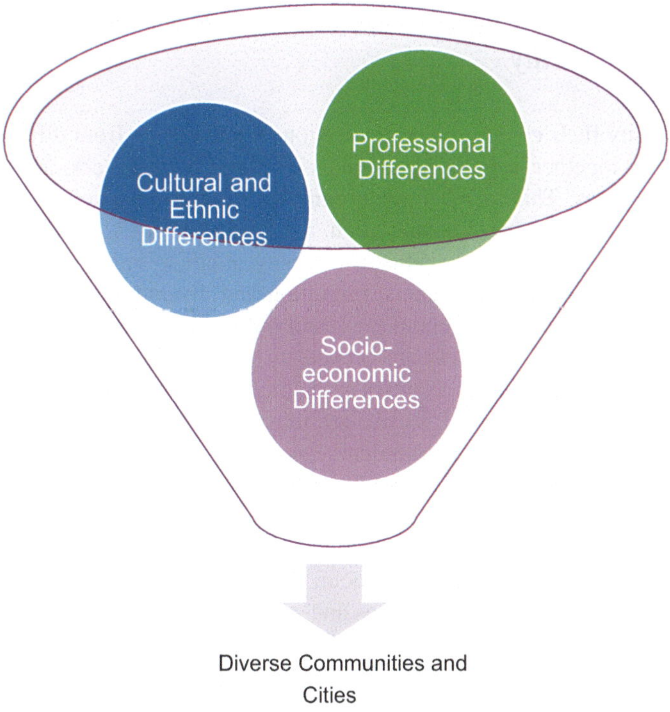

Fig. 1.1 Diversity encompasses various dimensions, including cultural, ethnic, socioeconomic, and professional differences

technological advancements and enhancing the city's competitiveness on a global scale.

Additionally, the book aims to explore how cities can implement policies and create inclusive spaces that welcome and support diverse populations. This includes strategies for affordable housing, equitable access to education and healthcare, and community-building initiatives that bridge cultural divides [26, 27, 36, 43, 49]—and some are beyond the scope or focus of the book. The examples provided demonstrate that by prioritizing diversity, cities can build more inclusive, equitable, and prosperous communities. In summary, this comprehensive compilation of global case studies emphasizes the necessary role of diversity in shaping vibrant urban landscapes. It serves as an invaluable resource for urban planners, policymakers, and anyone interested in creating cities that thrive on the richness of their diverse populations.

1.3.1 Diversity in Cities: An Embracing Engine for Urban Sustainability

Firstly, diversity fuels creativity and innovation. When people from different backgrounds come together, they bring unique viewpoints, experiences, and problem-solving approaches. This blend of ideas can lead to innovative solutions to urban challenges, such as housing, transportation, and sustainability. Unfortunately, there are misconceptions about the benefits of diversity in cities, some that even lead to blaming policies and agendas against migration and diversifying our urban environments [13]. However, throughout the history, diversity of backgrounds, knowledge, and socio-economic levels are proven to be important assets to cities and urban sustainability. For instance, cities with a high level of cultural diversity often become hubs for creative industries like art, music, and technology [1]. The cross-pollination of ideas in such environments can spark new products, services, and cultural expressions that enrich urban life.

Economically, diversity can act as a powerful engine for growth. Cities that embrace diversity tend to attract a wide range of talent, from highly skilled professionals to entrepreneurs and unskilled labour. This diverse workforce can adapt more readily to changing economic conditions and seize new opportunities. Furthermore, diverse cities are often more attractive to multinational companies seeking to tap into local and international markets. The varied consumer base in these cities can also stimulate demand for a wide array of goods and services, driving local businesses and contributing to economic dynamism.

Socially, diversity can enhance community resilience and cohesion. While it can present challenges, such as potential conflicts or misunderstandings, cities that actively promote inclusivity and equity can harness the benefits of diversity. Policies and initiatives that foster social integration, such as multicultural festivals, inclusive urban planning, and support for minority-owned businesses, can build stronger, more united communities. These efforts can reduce social tensions and create a sense of belonging among residents, regardless of their background.

Moreover, diverse cities are better positioned to tackle global issues like climate change and public health crises. The varied knowledge and practices from different cultures can offer valuable insights into sustainable living and innovative solutions to environmental challenges. For example, traditional ecological knowledge from indigenous communities can inform modern sustainable practices. Similarly, diverse healthcare practices can enhance public health strategies, making them more adaptable and effective in addressing the needs of different populations.

Education systems in diverse cities also stand to benefit. Schools with students from varied backgrounds can provide richer learning environments, exposing students to multiple perspectives and fostering critical thinking. Such educational settings prepare young people to thrive in an increasingly globalized world, where intercultural competence and empathy are essential skills. Therefore, embracing diversity in cities can lead to more inclusive governance [20, 41]. When city leadership reflects the diversity of its population, decision-making processes become more representative and equitable. This inclusivity ensures that the voices of all community members are heard and considered, leading to policies that better address the needs of the entire population. In sum, considering diversity as a catalyst in cities is not just a moral imperative but a strategic advantage (i.e., for urban sustainability, in particular). By embracing and leveraging the strengths of diverse populations, cities can enhance their creativity, economic vitality, social cohesion, and resilience, ultimately developing and advancing more vibrant and sustainable urban environments for all residents.

1.4 A Summary

This introductory chapter discusses how modern cities, as main hubs of economic activity and cultural integration, greatly benefit from 'diversity'. These urban areas, characterized by a mix of various cultures, significantly contribute to economic growth and a vibrant social fabric. Thus, the chapter summarized the crucial role diversity plays in fostering sustainable and resilient urban environments. Cultural diversity is closely linked to economic growth in cities. Diverse populations bring a wide array of skills and talents, fostering innovation and creating new business opportunities [40]. This diversity also makes cities more attractive to international businesses and investors, further boosting economic development.

Overall, the chapter provides a comprehensive analysis of the interplay between (urban) diversity and urban sustainability. It highlights the economic, social, and environmental benefits of diversity, offering practical insights for leveraging this asset in urban development. By embracing diversity, cities can build more inclusive, vibrant, and sustainable communities, ultimately creating a roadmap for the future of urban living. The following chapters are case study chapter contributions from scholars and practitioners around the world. They provide valuable recommendations and evidence-based examples of diversity as catalyst in cities and urban environments. This is crucial in the face of a rapidly globalizing era, where sustainability has become an oxymoron and diversity has become a vivified part of urban thinking. Thus, we hope the presented cases here provoke innovation and ideas for sustainable pathways that are local, context-specific, and people-centric. With all these in mind, a people-centric urban development is highlight recommended.

References

1. Allam Z (2020) On culture, technology and global cities. In: Cities and the digital revolution: aligning technology and humanity, pp 107–124
2. Allam Z (2021) On complexity, connectivity and autonomy in future cities. In: The rise of autonomous smart cities: technology, economic performance and climate resilience, pp 31–47
3. Allam Z, Cheshmehzangi A (2024) Sustainable futures and green new deals. Palgrave Macmillan, Germany and Singapore

4. Allam Z, Cheshmehzangi A, Jones DS (2023) Crafting and transforming colonial empires and landscapes. In: Climate and social justice: the political economy of urban resilience and mercantilism. Springer Nature Singapore, Singapore, pp 61–75
5. Allam Z, Sharma A, Cheshmehzangi A (2024) LiFE in the city: behavioural changes can drive urban sustainability goals. Cities 151:105163
6. Bairoch P (1988) Cities and economic development: from the dawn of history to the present. University of Chicago Press, Chicago
7. Bühler M, Hollenbach P, Köhler L, Armstrong R (2024) Unlocking resilience and sustainability with earth-based materials: a principled framework for urban transformation. Front Built Environ 10:1385116
8. Carey H, Florisson R, Giles L (2019) Skills, talent and diversity in the creative industries. In: Creative industries policy & evidence centre
9. Cheshmehzangi A (2023) Regenerating the local characteristics: the impacts on city life experience and identity of Shangri-La's old town. In: Mapping urban regeneration: city life experiences in Yunnan, China. Springer Nature Singapore, Singapore, pp 219–237
10. Cheshmehzangi A, Dawodu A (2019) Case study reviews: people, perspective and planning. In: Sustainable urban development in the age of climate change: people: the cure or curse, pp 69–131
11. Cheshmehzangi A, Dawodu A (2019) Stakeholder methods to climate change and sustainable development. In: Sustainable urban development in the age of climate change: people: the cure or curse, pp 23–68
12. Cheshmehzangi A, Li HMA (2020) Innovation through urban diversity and achieving comprehensive sustainable urbanism from a community-oriented approach. Curr Urban Stud 8(02):222
13. Cheshmehzangi A, Munday R (2022) From global to local: the case of migration and urban identity for regenerative city transformations. J Urban Regen Renew 15(3):308–324
14. Cheshmehzangi A, Tang T (2022) Chengdu-Chongqing city cluster: from an economic zone to a major economic circle in the Western China. In: China's city cluster development in the race to carbon neutrality. Springer Nature Singapore, Singapore, pp 77–92
15. Cheshmehzangi A, You N, Siri J, Birch E (2024) Harnessing urban innovation to unlock the sustainable development goals. Springer, Singapore
16. Chong SK, Bahrami M, Chen H, Balcisoy S, Bozkaya B (2020) Economic outcomes predicted by diversity in cities. EPJ Data Science 9(1):17
17. Clark G (2016) Global cities: a short history. Brookings Institution Press, Washington D.C.
18. Conti ME, Battaglia M, Calabrese M, Simone C (2021) Fostering sustainable cities through resilience thinking: The role of nature-based solutions (NBSs): lessons learned from two Italian case studies. Sustainability 13(22):12875
19. Cooney TM (2012) Entrepreneurship skills for growth-orientated businesses. In: Report for the workshop on 'skills development for SMEs and entrepreneurship, vol 28, pp 1–24
20. Das A, Dahiya B (2020) Towards inclusive urban governance and planning: emerging trends and future trajectories. In: New urban agenda in Asia-Pacific: governance for sustainable and inclusive cities, pp 353–384
21. Duxbury N (2012) Shifting strategies and contexts for culture in small city planning: interlinking quality of life, economic development, downtown vitality, and community sustainability. Cultural political economy of small cities. Routledge, New York, pp 161–178
22. Elias P (2020) Inclusive city, perspectives, challenges, and pathways. Sustainable cities and communities. Springer International Publishing, Cham, pp 290–300
23. Grant JL, Kronstal K (2010) The social dynamics of attracting talent in Halifax. The Canadian Geographer/Le Géographe canadien 54(3):347–365
24. Harris E, Franz A, O'Hara S (2023) Promoting social equity and building resilience through value-inclusive design. Buildings 13(8):2081
25. Isenberg D, Onyemah V (2016) Fostering scale up ecosystems for regional economic growth. In: Global entrepreneurship congress. Tagore LLC, Los Angeles, pp 71–97

26. Jackson A (2019) Cultural competency and racial inclusion. A research agenda for new urbanism. Edward Elgar Publishing, Northampton, pp 81–107
27. Jaramillo JE (1996) The community-building project: racial justice through class solidarity within communities of color. La Raza LJ 9:195
28. Kemeny T (2012) Cultural diversity, institutions, and urban economic performance. Environ Plan A 44(9):2134–2152
29. Levitt KP (2006) Reclaiming policy space for equitable economic development (revised text of a paper presented to The North South Institute, Ottawa, Canada, on January 19th). https://www.networkideas.org/featart/apr2006/Reclaiming_Policy_Space.pdf
30. Makalima M (2024) Addressing urban inequality through innovative social housing strategies: a comparative analysis of approaches and outcomes. Reg Bus Stud 16(1):39–50
31. Mega V (2013) Intergenerational cities embracing diversity and social justice. In: Quintessential cities, accountable to the future: sustainability, innovation and citizenship, pp 127–145
32. Monroy AM, Gars J, Matsumoto T, Crook J, Ahrend R, Schumann A (2020) Housing policies for sustainable and inclusive cities: How national governments can deliver affordable housing and compact urban development
33. Nijkamp P, Kourtit K (2013) The "new urban Europe": global challenges and local responses in the urban century. Eur Plan Stud 21(3):291–315
34. Ottaviano GI, Peri G (2005) Cities and cultures. J Urban Econ 58(2):304–337
35. Phan TT (2000) Tapestry of resilient lives: socio-cultural explorations of ten Vietnamese inner-city youths (Doctoral dissertation, University of British Columbia)
36. Pinkett RD (2002) Creating community connections: sociocultural constructionism and an asset-based approach to community technology and community building (Doctoral dissertation, Massachusetts Institute of Technology)
37. Ramos TB, Caeiro S, Disterheft A, Mascarenhas A, Deutz P, Spangenberg JH, Montano M, Olayide O, Sohal A (2020) Rethinking sustainability: questioning old perspectives and developing new ones. J Clean Prod 258:120769
38. Romice O, Rudlin D, AlWaer H, Greaves M, Thwaites K, Porta S (2022) Setting urban design as a specialised, evidence-led, coordinated education and profession. Proc Inst Civil Eng Urban Des Plan 175(4):179–198
39. Sax D, Manson C, Nesbitt L (2020) Governing for diversity: an exploration of practitioners' urban forest preferences and implications for equitable governance. Front Sustain Cities 2:572572
40. Smallbone D, Kitching J, Athayde R (2010) Ethnic diversity, entrepreneurship and competitiveness in a global city. Int Small Bus J 28(2):174–190
41. Suzuki H, Cervero R, Iuchi K (2013) Transforming cities with transit: transit and land-use integration for sustainable urban development. World Bank Publications, Washington D.C
42. Tasan-Kok T, Bolt GS, Plüss L, Schenkel W (2017) A handbook for governing hyper-diverse cities. Utrecht University, Utrecht
43. Thomas RL, Chiarelli-Helminiak CM, Ferraj B, Barrette K (2016) Building relationships and facilitating immigrant community integration: an evaluation of a cultural navigator program. Eval Program Plann 55:77–84
44. Tura N, Ojanen V (2022) Sustainability-oriented innovations in smart cities: a systematic review and emerging themes. Cities 126:103716
45. Turok I (2012) Cities, regions and competitiveness. Regional Competitiveness. Routledge, New York, pp 79–93
46. Wood P, Landry C (2008) The intercultural city: planning for diversity advantage. Routledge, New York
47. Zeidan I (2009) Cultural planning as a tool for urban development and cities regeneration (Doctoral dissertation, Birzeit University)
48. Zhuang ZC (2021) The negotiation of space and rights: Suburban planning with diversity. Urban Planning 6(2):113–126
49. Ziafati Bafarasat A, Cheshmehzangi A, Ankowska A (2023) A set of 99 healthy city indicators for application in urban planning and design. Sustain Dev 31(3):1978–1989

Chapter 2
Designing the Institutional Framework of Hybrid Special Economic Zones (HEZ)

Reza Mirzaei and Elnaz Lakizadeh

Abstract One of the challenging issues in today's era is the role of economic resilience of cities and countries against economic shocks in order to make decisions for the functioning of systems in the face of damage for economic growth. Today, in the literature of economic growth, special economic zones have become very important, which have been created in the form of large-scale industrial projects in the field of oil and gas and services in cities, with a long-term view and in close connection with macroeconomic fields. The activities of these areas are often focused on one dimension, which is either on the external dimension and only exports, or on the domestic dimension and the market. If the economic structure of cities should act as an ecosystem that includes a network of economic units related to each other; Because focusing on one dimension and not paying attention to diversity in the mix of industries affects the resilience of the city. Diversity is a critical factor in strengthening the city's capacity in line with national priorities that must be supported. This strengthens the strength of the city in various economic, social, cultural, and environmental dimensions against future shocks. For this purpose, the approach of hybrid special economic zones is defined, with diversity as its main shape. The purpose of creating these areas is to cluster areas in two dimensions: export-oriented and market-oriented. It is very important to pay attention to the policy loop of the hybrid special economic zone in the form of designing an effective institutional framework in two dimensions: sustainability-oriented and growth-oriented, which improves the resilience of cities. Iran's largest oil-related industries are located in Bushehr province under the name of "South Pars" which plays an important role in Iran's macro-economy. The export-oriented economy of this province is focused on oil and gas industries, in addition to its exports to foreign countries such as China, United Arab Emirates, India, France, Belgium, Germany, Poland, Norway, Russia, and Indonesia. Bushehr's economy is sensitive to foreign fluctuations and any changes in the demand of these countries, as well as the internal policies of the country and the province. This attitude, considering the country's special situation in the conditions of the tightening of sanctions and even the completion of oil development phases, can also negatively affect the city. Therefore, this research investigates

R. Mirzaei (✉) · E. Lakizadeh
Faculty of Architecture and Urban Planning, Tehran University of Art, Tehran, Iran
e-mail: Eng.RezaMirzaeii@gmail.com

G. Siew et al. (eds.), *Diversity as Catalyst: Economic Growth and Urban Resilience in Global Cityscapes*, Urban Sustainability, https://doi.org/10.1007/978-981-97-8776-0_2

the concepts of special hybrid economic zones and calculates its dimensions and components. Finally, an effective institutional framework for the internalization of the economy is presented along with the pursuit of the externalization approach, and the necessity of the country's focus on the components of resilience for economic growth is proven.

Keywords Hybrid special economic zones (HEZ) · Resilience · Economic growth · Institutional framework · Bushehr

2.1 Introduction

The economy forms the backbone of a country and plays a crucial role in its development. Therefore, economic growth is recognized as a potent tool that is essential for development, progress, and the enhancement of goods and services production. In the present era, one of the challenging issues revolves around cities and countries' ability to manage the impacts on their economic structures. Cities and countries face significant challenges when dealing with incoming shocks. In such cases, certain vulnerabilities are noticed in the city and the country, which reduce the quality of economic performance, society, and management in general. Today, resilience has become a central issue that encompasses political, institutional, structural, and macro and micro dimensions of a country's economy. In recent years, it has garnered the attention of numerous researchers, economists, and policymakers worldwide [8].

Following the great economic recession, the economic resilience approach emerged in the world of resilience literature. It provided a framework for policymakers and planners to navigate how to address adverse events and determine the appropriate responses in the context of resilience [13]. Therefore, the evidence confirms that the resilience of a region is related to its economy, performance, and structures. For this purpose, the researchers conceptualized economic resilience and demonstrated that resilience is not solely focused on recovering and returning to the state before the shock and crisis; it can also generate new opportunities and enhance the region's capacity. These new areas exhibit higher economic resilience compared to the previous areas. This event demonstrates an evolutionary process in economic resilience [9]. The core of economic resilience in this interpretation focuses on how the economy of a city or country can return to its state before facing a crisis and creating new opportunities. This definition of economic resilience highlights a transformative aspect that creates new opportunities for the economy and achieves structural transformation through the interaction and evolution of multi-scale institutions and industries [19].

Economic resilience is the combination of growth-enhancing policies with strategies aimed at estimating losses and determining recovery methods based on feasibility and effectiveness. Research shows that a successful strategy should include core measures to promote rapid economic growth while ensuring its sustainability. Static strategies refer to resiliencies that are possible when faced with a crisis by pursuing adaptation but disappear over time. They have limited effectiveness compared to dynamic resilience strategies. Dynamism and speed are the primary characteristics of dynamic economic resilience, which aim to repair and restore production capacity and accelerate growth [17].

Dynamic economic resilience strategies require catalysts to enhance their performance. Among the key and widely used tools in this field are the policies of industrial clusters, which are considered the driving engine of economic growth and development [18]. Special Economic Zones (SEZ) are a clear manifestation of industrial clusters that have a long history of coexisting with various economic and political institutions. They present themselves as the main strategies for investment, management, construction, and trade of goods and services [7]. Therefore, these regions are defined as catalysts that drive the development and reconstruction of the economic growth chain, creating a favorable compromise between the centralized government and the global market [4]. This definition likens Special Economic Zones (SEZs) to a catalyst in a chemical reaction that accelerates the economic resilience reaction [2].

By creating jobs, encouraging innovation, promoting industrial development and upgrading, attracting investments, diversifying the economy, and boosting export production, Special Economic Zones (SEZs) play a crucial role in enhancing economic resilience and fostering economic integration [5]. However, the main challenge is that Special Economic Zones (SEZs) often focus on a single sector, demonstrating a tendency towards one-sidedness by expanding SEZs to cater to the growth of industries that already exist. This expansion is occurring rapidly due to the political-economic nature of SEZs [6]. If diversity is the main factor contributing to the economic resilience of special economic zones, this resilience is manifested in a space that encompasses a combination of formal institutional structures, laws, agents, infrastructures, and services.

Hybrid special economic zones are defined as a combination of zones that include a set of industrial nodes and other services in an efficient manner [1].

The current literature on hybrid special economic zones (HEZ) in Iran is very limited. Moreover, the case of Iran is unique because it highlights political aspects of management that differ from Western approaches, which have distinct conceptions of networks of hybrid special economic zones. Therefore, the missing link for establishing HEZs in Iran and South Pars is the presence of an institutional framework, which is essential for the successful implementation of HEZs in industrial clusters. The implementation of the institutional framework is a crucial process to lay the groundwork for the economic resilience of Bushehr and South Pars.

Special Economic Zones (SEZs) have emerged in Bushehr province and the South Pars oil and gas field, strategically located around existing resources, particularly oil and gas. They are mostly created in areas that have the potential for oil and gas exploration or the development of related industries. However, Special Economic Zones (SEZs) in Iran have generally been established at the behest of oil policymakers, who are driven by the goal of addressing political-economic challenges by relying on oil. Unfortunately, these zones often lack integrated strategies and policies for local development and urban planning. As a result, while some projects have been developed as examples in the documents and laws of the South Pars Special Economic Zone, most of them do not prioritize resilience and remain at the descriptive level. Focusing solely on the upstream and downstream industries dependent on oil in South Pars makes this region vulnerable to damages during a crisis, the negative effects of which are evident in the country and the province (see Fig. 2.1).

The purpose of this study is to establish an institutional framework for utilizing HEZs to enhance the resilience of the South Pars Special Economic Zone through a practical approach, drawing on relevant articles and documents. The study aims to address a fundamental question: What is the most effective institutional framework for the hybrid special economic zone of South Pars? To answer this question and achieve the mentioned goal, the concepts, dimensions, and components of special hybrid economic zones have been identified. These zones are primarily defined by policies aimed at enhancing economic resilience. In general, this study is related to the overall theme of the book, which examines diversity as a catalyst for economic growth and urban resilience in global urban spaces.

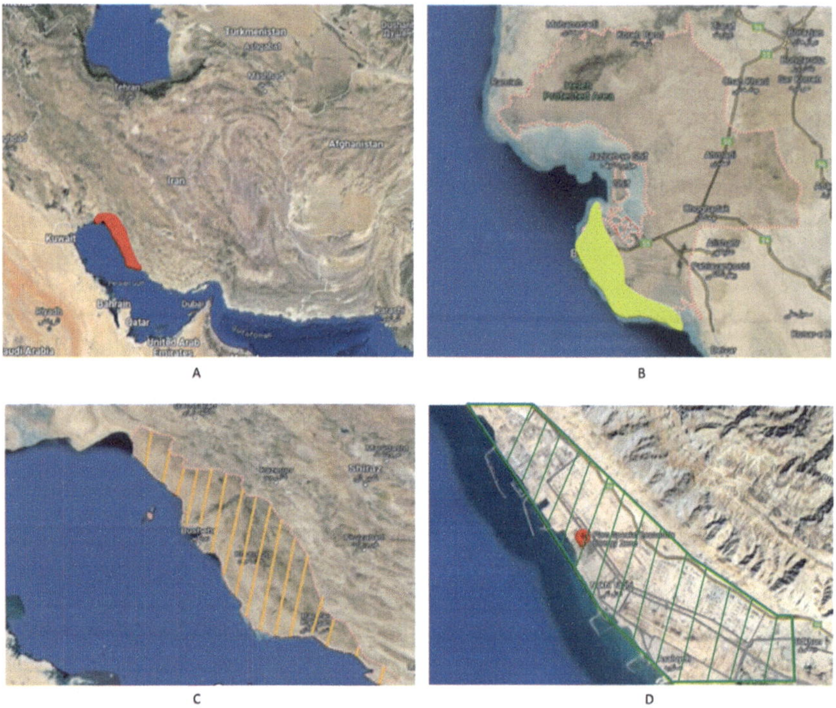

Fig. 2.1 Introductory image of the South Pars Special Economic Zone, depicting **a** the geographical location of Bushehr province in Iran, **b** the geographical location of Bushehr city in Bushehr province, **c** the geographical location of Pars Special Economic Zone, and **d** the South Pars economic visa area (*Image Source* Drawing by the Authors)

2.2 Background Knowledge

2.2.1 Definitions of Hybrid Economic Zones

The complexity of the global economic system, involving cities and countries, necessitates the utilization of innovative tools and catalysts to enhance countries' capacity to address economic crises and sustain economic growth. The twenty-first century can be described as the peak of innovative activities. A good example is the hybrid economic zones where diversity is their main characteristic. Studies have shown that the conceptualization of hybrid economic zones is defined based on the functional scale, type of management structure, and activity nodes, which will be discussed further.

Before the introduction of the hybrid economic zone approach by the World Bank in 2019, special economic zones were focused solely on one dimension, operating as single-function entities. Additionally, the management system was either publicly or privately administered. With the emergence of HEZs, diversity became the main pillar of economic zones, and the concept of hybrid economic zones or combined economic zones rose to prominence. In the economic regions with general management, the institutional framework and system of regional administration were implemented by the regional administration, through which the space users were selected. In this structure, the central and local governments delegate their powers, especially in the field of supervision, to the regional administration, which has more limited operational independence compared to the private model. Privately managed economic zones have private developers who enjoy more operational autonomy compared to the public model. In this structure, the responsibility for concluding investments, arranging land rent, costs, and other operations lies with the private sector. Hybrid economic zones are a combination of public and private management models. HEZs maintain relatively broad autonomy in the operation of the zones. In this framework, various users are drawn to the area, and the status of these users is regulated by contracts. Hybrid economic zones offer greater flexibility compared to previous models, enabling policymakers to tailor various types of Special Economic Zones (SEZs) according to the existing activities in the region, the region's potential, and specific projects. In this structure, unlike the previous models where the regions were enclosed in a fence and the contribution of local governments and local communities was limited, the possibility of local government participation is also provided (World Bank 2019).

Hybrid economic zones are areas that are generally divided into two parts based on Special Economic Zones (SEZ) and non-SEZ areas interacting with each other. These zones have emerged as a combination of various industrial clusters on both cross-border and national scales. In the concept of national hybrid regions, a single growth pole transforms into a hybrid region comprising a set of functional nodes. This model utilizes the generated effects on a larger scale. On the scale of cross-border hybrid regions, the revitalization of former economic regions occurs through the expansion of the number and types of new programs [1].

2.2.2 The Evolution of Special Economic Zones

The origin of the emergence of special economic zones can be traced back to free ports centuries ago when merchants operating on ships transported and exported goods without the intervention of local authorities. In the modern era, starting in the 1960s, special zones began to emerge in proximity to border corridors, seaports, and airports. The World Bank's 2017 report on the operational review of the effects of special economic zones showed that the first modern zone in 1937, which catalyzed the development of SEZs, was created at the East River location in Brooklyn, at the New York Navy Yard. The first free trade zone in Europe was established in 1959 in Shannon, Ireland. The expansion and development of these regions intensified with the implementation of export-oriented strategies starting in the 1970s. The wave of changes accelerated significantly in the late 1990s and early 2000s, coinciding with the rise of international production and the rapid expansion of global value chains (GVCs). This period also saw the creation of new types of SEZs, as governments responded to economic competition by establishing more SEZs. Therefore, the evolution of special economic zones can be traced through six generations, beginning with industrial special economic zones and culminating in hybrid special economic zones. The emphasis on HEZs in the current economic crisis, along with the inefficiency of unilateral economies, highlights the necessity of pursuing approaches aimed at enhancing economic resilience in cities worldwide (see Table 2.1).

The negative effects of the economic crisis are evident in other examples of special economic zones, where a focus on unilateral goals has led to a loss of economic resilience. This event prompted economists and planners, especially those in charge of economic affairs at the UN Trade and Development Conference in 2021, to discuss the economic resilience of special economic zones. It should be noted that hybrid economic zones were first mentioned in the World Bank report in 2019. The purpose of these zones is to provide a combination of functional nodes with public–private management. The existing diversity acts as a catalyst that, with a multifaceted view, on the one hand, fosters economic growth, and on the other hand, enhances economic resilience in both macro and micro dimensions, directly and indirectly, in market and non-market sectors. One of the main features of these regions is the maintenance of economic balance in the region.

2.3 Methodology

This study is conducted with a practical approach. In the present study conducted in Iran, Bushehr Province, Pars, and South Pars Special Economic Zone were selected as the study sample. Information was collected using field and library-document methods.

Table 2.1 The evolution of special economic zones

Types of special economic zones	The reason for its origin	Objectives	Potential effects on economic resilience
Special economic zones (SEZs)	The need for organized export areas	Exporting overseas	*Negative effects:* lack of diversity in jobs, income, working classes, knowledge and skills, infrastructure, business interruption, production losses, supply chain disruptions, and violations of women's labor rights in favor of men, with women having a small share *Positive effects:* The official establishment of special economic zones, the transfer of knowledge related to the common industry, specialized expertise, and innovative practices within the common industry, the concentration of common industries in one location leading to technology spillover, and the boost in national-level exports
Export processing zone (EPZ)	The need for a catalyst in rapid growth, economic transformation, and earning	Facilitating foreign investment, earning foreign currency, and creating a privatized and secure space	*Negative effects:* decrease in the share of the local economy and individual industries, an increase in average distance and commuting time, income inequality, disparities in employment opportunities, lack of job diversity, lack of infrastructure diversity, and limited linkage effects with the domestic and local economy *Positive effects:* Limited attention to the expansion of economic zones
Environmental industrial parks (EIP)	Global climate change resulting from the rapid expansion of Special Economic Zones (SEZs)	Ensuring environmental, social, economic, and managerial sustainability involves fulfilling national and international commitments	*Negative effects:* lack of diversification in industries *Positive effects:* creating a green supply chain, improving resource management, and enhancing environmental resilience
Foreign trade zones (FTZ)	The Need for Free Trade	Focusing on service sectors in free trade to improve the foreign investment environment	*Negative effects:* reducing the share of the local economy in individual industries and increasing income inequality *Positive effects:* expansion of diversity and broadening of attention in the region

(continued)

Table 2.1 (continued)

Types of special economic zones	The reason for its origin	Objectives	Potential effects on economic resilience
Digital free trade zone (DFTZ)	The Need for Electronic Commerce Following the Revolution in Information and Communication Technology	Electronic export and import services make areas more efficient and environmentally friendly, creating favorable conditions for small and medium-sized companies to enter the business market	*Negative effects:* reducing the share of the local economy and individual industries, solely focusing on the electronic dimension, and promoting unilateralism *Positive effects:* Expanding the field of vision towards the integration of industry and electronics, creating job diversity in the information and communication technology sector
Hybrid economic zones (HEZ)	The a need to pursue a hybrid approach in special economic zones for economic resilience	Pursuing public–private management involves offering a variety of services through a series of functional nodes	*Negative effects:* price increase in the region *Positive effects:* diversity of industries, diversity of employment and self-employment, diversity of income, diversity of the workforce in the form of paying attention to the necessity of guaranteeing women's labor rights, increasing the share of local capital, increasing knowledge and skills, and increasing infrastructure diversity, Reducing the risk of consumption and property and welfare, economic spillover, favorable linkage effects with national and local domestic economy, local neighborhood of domestic and foreign companies, transfer and spillover of knowledge and technology, facilitation of regional performance, and economic growth

Source Authors' studies adapted from Kahsani [11], World Bank (2019), Zeng [21] and Joseph [10]

2.4 Hybrid Economic Zones

The history of oil and gas in the special economic zones of southern Iran is a part of the common heritage of humanity, recognized as a foundation for economic growth during the modernization era. However, up to this point, there has been no comprehensive understanding of the effects that the one-sided view and sole reliance on oil and gas in Iran's special economic zones have had on economic resilience. Although this study has a practical dimension and deals with presenting the general discourse of the country's special economic zones, its main focus is on the institutional framework necessary to transform the single-functional special economic-industrial zone of South Pars into a hybrid economic zone. This hybrid zone would provide a combination of functional scale, type of management structure, and activity nodes.

The first section argues that the need for Special Economic Zones (SEZs) to be resilient stems from the fact that now, more than at any other time in history, crises and events threaten cities and countries in the form of shocks. The ability of these regions to resist, recover, and move forward following these events depends on their economic resilience. On the other hand, economic resilience depends on a variety of functions, management structures, and activity nodes. The second section discusses the factors that drive investment in hybrid economic zones. The third section argues that Iran's public discourse has not been able to provide a satisfactory framework for presenting mixed economic zones, and the existing economic zones are summarized as a combination of industry and residence. The term "satisfactory" refers to the concept that integrated thinking and a comprehensive approach to economic resilience, rather than relying solely on one dimension such as oil and gas, have not been institutionalized in the country's planning and policy-making, especially in Bushehr provincial planning documents. The last part argues that preparing the institutional framework of HEZs can ensure attention to economic resilience policies in the structure of the country's economic zones, with a focus on South Pars.

South Pars 1 and 2 are located in Bushehr province, near Asaluyeh and Kangan. These fields supply the majority of the country's oil, gas, and materials required by other industries. Examining the policies in this area reveals the one-dimensional perspective of the country and Bushehr on expanding this special economic-industrial zone through the development of upstream and downstream oil and gas industries. Two points are important for this study. The importance of policies and financial incentives for the creation of HEZ should be addressed. Secondly, the importance of establishing a dedicated institutional framework for hybrid economic zones as a tool for building a resilient economy should be addressed.

2.4.1 Hybrid Economic Zones Driving Economic Resilience

The economic resilience literature is a complex and evolving field characterized by significant challenges related to the uncertain and emerging nature of threats, the speed of change, the level of complexity of threats and changes, policy-making, political pressures, and governance systems. In addition, concepts play an important role in the development of economic resilience policies. The existing conceptualizations are accompanied by a wide range of criticisms. The most significant criticism is that these concepts often aim to revert to past ways of life, which are not well aligned with the other goals of the Special Economic Zones policy. Additionally, they take a short-term view in terms of time.

The research approach here should be recognized and valued as a hybrid approach. This approach can have a positive impact on decision-making, policy formulation, enhancing economic resilience, and supporting diverse regions. This can be achieved through the pursuit of an effective institutional framework, diversity in activity nodes, and performance scale. Finally, it can be said that the criteria for economic resilience in Special Economic Zones (SEZs) are the same components and indicators found in hybrid economic zones. The following discussion presents an integrated view of economic resilience and its driving force, the HEZ.

Today, the economic crisis and its effects are among the topics that have received a lot of attention. The important point is the lack of prior knowledge regarding crises and damages that can affect economic regions and threaten the resilience of the economy. Studies have shown that limited management capacity, financial and human capital, information, and adherence to fundamental internal knowledge are deeply ingrained in the traditional approach to these areas. This results in these areas consistently maintaining a low level of economic resilience and being highly vulnerable to damage during crises of varying scales. On the one hand, the lack of strategy, attention to modern innovations, excessive focus on profitability from the production unit, governance culture, and government structure, limited time, financial and human resources, overreaction to external influences, lack of management skills within small and medium companies, entrepreneurial orientation, and the search for opportunities also contribute to making economic resilience in regions a challenging issue for policymakers. At the same time, the economic survival of special economic zones relies on paying attention to the above principles [16].

In the political dimension, government in public administration operates within various economic-political systems and encompasses diverse duties and powers. Government in countries with a centralized structure plays an essential role in the development of economic resilience of Special Economic Zones (SEZs). Meanwhile, the government's role is influenced by various political, diplomatic, economic, and socio-cultural factors. This set of factors affects crisis management systems and ensures technical support, damage control, and risk management.

Therefore, the governing structure must ensure the existence of a crisis management system, its development, and the establishment of processes and policies that govern risks through the design of laws and regulations, decision-making procedures, legislation, and policy implementation mechanisms. It should also promote the necessary awareness and knowledge at various national and local levels [12].

These factors, which were discussed earlier, demonstrate the material, social, and political inequality within an economy. This inequality leads to instability in the economy and erodes trust in the government among communities. Therefore, dealing with different systems is a prerequisite for creating new economic realities to address the challenges posed by crisis barriers. Therefore, transitioning from traditional strategies to emerging strategies is crucial. This transition guides policymakers. Additionally, it is essential to invest in diverse infrastructures, measure the effectiveness of resilience, and establish and promote system evaluation mechanisms. In this perspective, the alphabetic language of economic resilience involves reducing damages, integrating policies and measures, examining and evaluating possible alternative options, and accepting the necessity of transformation. This provides a clearer framework for application to enhance economic resilience and address crises and damages [3].

Therefore, there is a need to change the approach. In today's era, the economic environment of countries is being threatened and damaged by various factors. In response to the need to adapt to these changes and potential crises, resilience is recognized as a fundamental pillar in economic regions. Resilience, while it involves a return to the previously known way and the past situation, also encompasses moving forward. For this reason, it is recognized as a powerful tool for overcoming crises and advancing to enhance the economic strength of cities and countries. This underscores the importance of focusing on emerging economies [20].

Emerging economies and their corresponding strategies have the potential to steer the fluctuations, crises, and damages of the economy in the international, national, and local arenas towards resilience. In recent years, the intensification of economic instability, the unilateral reliance of countries on international markets, the interconnectedness of global economies, and the influence of each other's economic consequences in the form of exposure to external shocks, the lack of diversity as an effective catalyst for economic growth, and more importantly, weak and ineffective institutional frameworks, which are caused by many factors, especially political causes, have created tension in the economies of countries and have increased the importance of economic resilience as a promising solution and a guiding light [14]. Therefore, there is a need for a stimulus that incorporates these factors. Hybrid Economic Zones (HEZ) fulfill this need.

As a result, HEZs as catalysts for economic growth require integrated thinking and a comprehensive approach to economic resilience in planning and policymaking. This opportunity can create a new mechanism to not only maintain the initial dynamism of special regions but also enhance their capacity to respond effectively in times of crisis. Therefore, economic resilience plays a crucial role in supporting hybrid economic zone management scenarios, addressing key issues such as reducing vulnerabilities and enhancing the stability of these zones. On the other hand, Hybrid Economic Zones (HEZs) are recognized as significant catalysts for boosting economic resilience by enabling a return to previous economic conditions while also advancing into new economic territories. This is achieved by fostering diversity in services, moving away from dependence on a single industry, and promoting the growth of associated industries within a robust institutional framework. Special Economic Zones (SEZs) enable countries to not only integrate with the global economy but also to develop a robust domestic economy that can withstand external economic shocks (see Table 2.2).

The economy is often referred to as the growth engine of a country. The continuity of financial and institutional dimensions is reflected in the economic growth of countries. Therefore, investment and its drivers are crucial for the economic resilience of a country and understanding them can serve as a valuable guide for HEZs. In the following, investment incentives are discussed.

Table 2.2 Dimensions and components of resilience and hybrid economic zones

Hybrid economic zones		Economic resilience	
Functional scale	International, national, local	Structural foundations	Economic diversity, human capital index, technology adoption, labor market flexibility, entrepreneurial activities and business dynamism, poverty index, industrial diversity, new physical capital/ infrastructure (asset), and promotion of new job opportunities
Activity nodes	Services and infrastructure	Institutional structure	Political stability, rule of law, voice and accountability, efficient government, control of corruption, support by a broad political consensus
Management-institutional structure	Organizational quality, public, private or public–private partnership, centralized/ decentralized implementation, policymaking, and legislation at the country level	Money space	Financial development, financial stability index, non-current loans, capital adequacy ratio, liquidity ratio
Financial space	Investment, financial packages, income, gross domestic product, tax exemptions	Financial space	Debt to GDP ratio, budget deficit to GDP ratio, service debt to income, debt to tax revenue, labor income, investment income, business income, transfer payments, income inequality, inflation, demand Domestic, financial stability, and wages

(continued)

Table 2.2 (continued)

Hybrid economic zones		Economic resilience	
Commercial space	Trade diversification, current account/ GDP, investment position, access to source and destination markets, integrated labor markets	Commercial space	Export diversification, current account/gross domestic product, reserve/ short-term debt, net international/investment position, policies to diversify the economy and pursue green projects
Social Space	Human capital, population density, and the right of participation of all classes in the division of labor	Social Space	Age distribution, educational attainment, health, labor mobility, creative class, population size and density, settlement density, highly skilled and educated workforce

Source Authors' studies

2.4.2 Investment Drivers Toward Hybrid Economic Zones

As mentioned in the previous sections, diversity is a prerequisite for economic resilience and is a key element that the government must monitor to diversify the economy. In this context, according to Marshall's theory and model of economic density, industrial accumulation is formed in the region. A mechanism of self-reinforcing agglomeration is created through the clustering of SEZ-based and non-SEZ-based industries into HEZ, where they interact with each other. On the other hand, the aggregation of producers, suppliers, and consumers forms an ecosystem of economic efficiency, creating mutual dependence. This dynamic fosters regional development and economic spillover into local economies and surrounding areas, laying the foundation for economic growth. The financial environment is at the core of any health, environment, and safety (HEZ) policy aiming to establish an efficient financial framework for the country. Therefore, the characteristics of HEZ affect the economic performance of the entire region in terms of internal economy, including economic spillover at the national and local levels, external economy, such as the level of international exports, and planning for incentives and investments. HEZs are considered by policymakers as a tool for dynamism and an important part of development strategies that promote long-term economic growth through spillovers.

In response to the question of what drives investment in special economic zones, the following can be addressed by examining the relationship between hybrid economic zones and the domestic economy, foreign economy, as well as incentives and investments.

2.4.2.1 HEZ and the Domestic Economy

One of the main factors driving investment in special economic zones and increasing countries' interest in hybrid economic zones is the relationship between the HEZ and a country's domestic economy. Domestic economy refers to the economy at national and local levels.

The local economy operates as a protectionist system that, with a justice-oriented approach on a small scale and in the local dimension, prioritizes placing citizens at the center of power and wealth production, significantly impacting people's lives. The local economy plays a strategic role in both the economy and politics, fostering trust in the government. At first glance, the goal of a local economy is to empower local communities, especially low-income and vulnerable groups, to have the capacity to produce according to their needs and safeguard their production capabilities. The characteristic of this approach to the local economy is to empower local communities to participate in local financial circulation and achieve a more equitable distribution of wealth. Secondly, another goal of exporting is to meet surplus production needs, which the local economy aims to achieve to broaden its participation in financial circulation. HEZ strengthens the local economy by improving local production standards, increasing the share of local communities in capital circulation, and diversifying jobs and services. The reflection of this relationship is evident in the establishment of a stable and resilient society from a socio-economic perspective, where local residents have economic autonomy and greater oversight of their local resources. Thirdly, in this way, companies are enabled to engage with the local economy and demonstrate the spillover theory. This theory suggests that spillovers happen within the company and are reflected in increased local employment, innovation, and growth. Institutional frameworks, structural foundations, laws, regulations, infrastructure, financial support, and innovations shape the relationship between HEZ and local markets. The spillover of the local economy to higher national levels contributes to the economic growth of the entire region. The national and local economies are mutually interacting through economic spillover, a topic that will be further discussed.

The national economy is considered the embodiment of a nation, shaped by its geography, location, unique laws, and political structure. The countries of the world pursue various objectives to secure their economic stability. The relationship between HEZ and national markets can be gauged by indicators such as gross domestic product, trade, investment, monetary policies, labor market, production, inflation, distribution, and consumption of goods and services. Hybrid economic zones contribute to national economic growth by bridging infrastructural gaps, enhancing diversity within the framework of convergence criteria, and laying the foundation for building a resilient national economy. On the other hand, the spillover theory in other fields of the economy, such as finance, development, knowledge, technology, and skills, in the national economy shows that spillovers occur either horizontally within an industry or vertically between industries. Meanwhile, the economic spillover effect of HEZ spreads vertically and manifests as positive impacts of an event based on HEZ in the national economy on other economies, particularly the local economy. The national economy and the foreign economy, which relies on exports, have a mutual relationship that will be further discussed.

2.4.2.2 HEZ and the Foreign Economy

Another significant factor that attracts investment to special economic zones and increases countries' interest in hybrid economic zones is the connection between the HEZ and a country's foreign economy. Exports are at the center of attention in the foreign economy. The relationship between HEZ and international markets can be observed through various indicators such as trade (country growth, export share, market density index, market penetration index, import share, etc.), tariff (share of tariff lines, share of tax tariff lines, import tax, etc.), and development (export of business services, gross domestic product, export of information and communication technology services, insurance and financial services, annual growth percentage of export of goods and services, etc.). By enhancing the local economy, HEZ can bolster the role of this economic sector in exports and lay the groundwork for establishing a resilient national economy. In this type of economy, the interconnected national economies of countries play a crucial role. The important point in this section is the effects of economic spillover. In explaining this issue, it can be stated that economies are interconnected globally. Consequently, spillovers on a global scale can have numerous positive and negative effects, leading a country to experience shock and damage due to economic challenges faced by another country. Meanwhile, HEZ can mitigate the effects of damage by diversifying industries, businesses, and services.

2.4.2.3 HEZ and Incentives for Investment

Another factor that attracts investment to special economic zones and increases countries' interest in hybrid economic zones is the connection between HEZ and the diversity of incentives and types of joint investments. HEZ policies include financial incentives such as subsidies, financial support, tax credits, tax exemptions, and reductions in tax and tariffs. Additionally, they consider special non-financial incentives to attract companies to these areas and achieve greater economic efficiency in the development of the region. On the one hand, the potential for joint investment and investment diversification is facilitated through hybrid economic zones. In this type of investment, foreign companies, domestic companies, and cooperatives, especially, collaborate in the local sector. This collaboration enables cross-border, intra-border, and national and local business cooperation. HEZ supports investment diversification and is a strong pillar of economic resilience.

As a result, it can be said that economically, countries with a robust financial and commercial environment have a greater ability to revive their economy without compromising financial resilience. They also can maintain their status in the face of internal or external shocks and crises. Finally, a country with a robust financial environment that can withstand shocks without triggering inflation is known for its favorable economic resilience. Another crucial aspect in economic regions is institutional factors or frameworks that play a significant role in economic resilience. In the following, the existing framework will be discussed first, followed by examining the effective institutional framework.

2.4.3 The Existing Framework: What Does It Offer?

Examination of the South Pars gas field reveals its strategic significance. The South Pars/North Dome Gas field is the largest gas source in the world. It is situated on the common border line between Iran and Qatar in the Persian Gulf. The Qatar side commenced its production in 1369 AH (1990 AD). In Iran, the South Pars oil and gas fields were discovered in 1349 AH (1970 AD), exploited in 1367 AH (1988), and production commenced in 1379 AH. This shows a ten-year lag in production. Examining the Land Development Document of Bushehr Province approved in 1394 AH (2015 AD) as an available document studying its policies and strategies, and examining the prospect of South Pars fields shows that the goal of creating a special Pars region is the development of South Pars gas phases and compensation for the country's lag in exploitation due to the Iraq war and damage to equipment. The development of petrochemical industries aims to create more added value from the South Pars gas reserve. The strategy of developing and increasing production to increase the number of phases is one of the main goals to compensate for the country's backwardness and to meet 70% of the country's gas needs in the South Pars region. This shows that the war crisis in this region in the 1360s (1980s) had numerous effects.

The war damaged and shocked the economic resilience, paving the way for the country's backwardness. Examining the current situation reveals that after the war, policymakers aimed to restore South Pars to its former state and enhance production. This effort has been sustained by a focus on advancing upstream industries and supplying oil and gas-related products and services. The special economic zone of Pars Energy includes two parts: North Pars and South Pars. Today, there are 29 phases in South Pars, which are divided into two sections, Pars 1 and Pars 2. The operational site of Pars 1 is situated in Asaluyeh, covering an area of 14 thousand hectares and comprising 21 phases. The operational location of Pars 2 is in Kangan, covering an area of 16,000 hectares and consisting of 8 phases. The economic experts of the country, in 1402 AH, analyzed the role of the production of South Pars phases. They stated that the production of each phase contributes 1% to the country's gross domestic product and plays a significant role in the financial and commercial environment of the country at the macro level.

The public discourse in Iran regarding the governance and administration of the country indicates that the governing structure is centralized, with planning, decision-making, and policy-making being conducted from the top down. Key actors in this area include the Ministry of Oil, the National Iranian Oil Company, and the South Pars Special Economic Zone Organization. Special economic zones that frequently align with industrial performance in the oil and gas sector are not immune to this principle and are significantly impacted by the spillover effects resulting from the political decisions of the country's governing body. This is due to the fact that the foundation of Iran's economy relies on oil and gas. The South Pars Special Economic Zone serves the purpose of economic development and plays a crucial role in advancing the country's overall development. Therefore, the development strategy for this region has been outlined in the approved plans of Faradast, focusing on the physical dimensions (land requirements) and performance (production materials) of these regions.

Examining the goods, services, and variety of industries shows that the spillover effects from South Pars occur horizontally and within the industry. The interests, goals, and priorities of the country's governance structure are centered on the advancement of upstream, downstream, and integrated industries associated with oil and gas. Therefore, knowledge spillover also occurs between these industries. Emphasis on the development, expansion, and investment in downstream oil, gas, and petrochemical industries; focus on creating industrial employment opportunities; emphasis on economic development and increasing domestic investment. Additionally, emphasis is placed on developing the network and industrial and service chain of the upstream oil and gas industries in the province and expanding oil industry services in the region. These policies are jointly pursued by stakeholders in managing South Pars.

The surveys above demonstrate that South Pars, as a Special Economic Zone (SEZ), aims to diversify its oil and gas-related industries. This diversification leads to knowledge spillovers and attracts labor to these sectors. The local economy in this region has not been adequately addressed, indicating that the contributions of local communities and capital are not being considered. As a result, most of the profits from goods and services benefit the macroeconomy. South Pars heavily relies on exports. In the event of a crisis in the target cities, it would experience a significant shock, which could have repercussions at the macroeconomic level of the country. Hence, the South Pars economy lacks resilience.

2.4.4 The New Framework: What Can It Offer?

Frameworks are comprehensive tools that provide a structured plan in a specific area. The development and use of frameworks are the most general forms of analysis that illustrate the elements and their relationships. In this format, a set of essential variables needed for analyzing institutional-organizational arrangements is expanded. Sometimes, the lack of an efficient institutional framework can reduce the effectiveness of resilience measures in practice because laws and regulations are not effective without a coherent structure. In economic zones, having an institutional framework to implement hybrid economic zones and identify various structural variables is crucial for creating a resilient economic zone.

In the first step, the main structure of the framework is presented. It branches off from the system based on the indicators of resilient hybrid economic zones and focuses on the institutional nature of the framework. This type of classification enables policymakers and planners to effectively implement economic zones.

In the second step, after determining the effective factors in an efficient institutional framework for resilient HEZ, the main dimensions are defined. Therefore, four dimensions are identified for the institutional framework that demonstrates how HEZs should be managed to achieve economic resilience. For the institutional framework of HEZ, these dimensions include characteristics of actors, institutions, public goods or services, and the society under investigation. The integration within the institutional framework encompasses a broad spectrum of institutions, rules, actions, operations, and services [15].

In the third step, after determining the dimensions of the HEZ institutional framework for each, specific target groups are defined. For the next, the characteristics of the target group institution include risk management, economic management, transformation management, public policy management, control and supervision management, asset management, and government organization design management. These elements are crucial in the management of hybrid economic zones, ensuring that the spillover effects of macro policies extend into these areas. For the actors' dimension, the target groups include public–private actors. Next, the characteristics of goods and services for the target group include service management, which forms the main pillar in hybrid economic areas. In the context of societal characteristics, the collaborative role of local funds is discussed. In the first stage, this framework provides strong support for the creation of a Hybrid Economic Zone (HEZ), which in turn becomes a significant driver for the economic resilience of South Pars, ensuring a robust economy. This framework will be discussed in the next section.

2.5 Discussion and Implications

The HEZ institutional framework provides a multi-layer conceptual map that includes institutional characteristics, actor characteristics, goods and services characteristics, and community characteristics. Next, in Fig. 2.2, the dimensions and target groups of each are explained (see Fig. 2.2).

2.5.1 Institutional Features

Public Policy Management: Policies are tools that outline a range of acceptable government actions, activities, and objectives by government institutions. They provide a framework for decision-making and offer insight into how organizations operate in the present and future. Public policy management plays a significant role

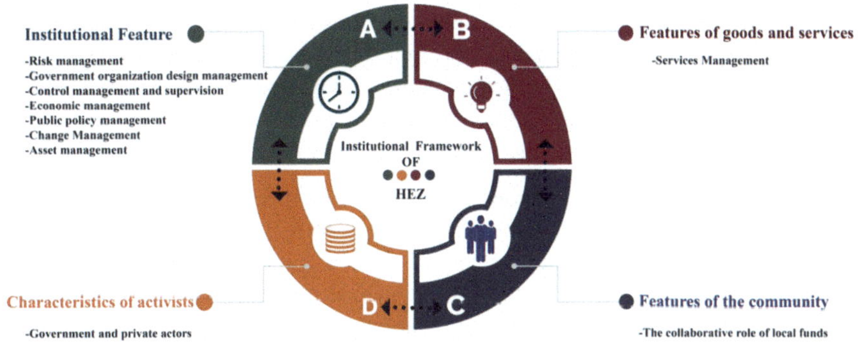

Fig. 2.2 Introductory image of the HEZ institutional framework, which includes **a** institutional characteristics, **b** characteristics of goods and services, **c** characteristics of society, and **d** characteristics of actors and activists in the field (*Figure source* Drawing by the Authors)

in identifying problems as issues in various eras. Making a policy on the economic resilience of HEZs today involves recognizing the problems and challenges faced by previous SEZs. Relying solely on an industrial cluster and absorbing negative effects in the face of shocks highlights the need to develop a comprehensive vision for the establishment of HEZs as a combination of industrial and non-industrial clusters. Therefore, a multi-sectoral, multi-objective, and multi-space policy is important for the implementation of HEZ.

Risk Management: Identifying possible risks and threats that a system faces reveals the necessity of risk management. The intersection between risk management and economic resilience lies in addressing the existing one-sidedness of special economic zones and concentrating on a specific dimension. Determining goals to assess a country's willingness and ability to achieve them, identifying potential risks to anticipate negative incidents, evaluating risks and their impact intensity to prioritize them, defining responses and solutions for each risk to establish necessary measures at different risk levels, and monitoring the effectiveness of the response are steps involved in the risk management of Special Economic Zones (SEZs). HEZ is defined as a response to the risks and crises facing SEZs, which can be transformed.

Change Management: Change is one of the main pillars of success in the institutional system. It is an inevitable reality in the process of transitioning from a frozen state to a new, compatible state, and then re-freezing in a compatible manner. Inconsistency can heighten vulnerability in crises. The intersection between transformation management and economic resilience in special economic regions lies in the recognition of the necessity for change and transformation due to inefficiencies in the current adaptation. This situation reflects a lack of diversity in industries and services, insufficient focus on the local economy, and overreliance on a single industry. Change management addresses the legal aspects of attending to this need. Therefore, it is necessary to devise a solution to steer the necessary changes and transformations towards a constructive and proper adaptation, incorporating essential modifications to foster a resilient economy. HEZs are recognized as catalysts and tools for seeking and establishing new adaptations that form the foundation for fundamental change and transformation, as well as for achieving desired outcomes in economic sectors. Economic resilience is ensured during this transition and transformation, which can result in a new organizational structure.

Design Management of Government Organizations: optimizing and ensuring effective performance is at the core of management for organizations to achieve their goals. In the field of managing special economic zones in South Pars, where the focus has traditionally been on one-dimensional performance related to oil and gas production and exploitation, there is a crucial need to re-engineer the primary management approach. This re-engineering process leads to organizational changes and the establishment of a dedicated entity to oversee HEZ management. The organization and planning of hybrid economic zones according to their specific needs have resulted in effective management, ensuring a resilient economy.

Economic Management: The research illustrated a strong correlation between special economic zones, economic resilience, and the economy in general. Most of this correlation was observed in the economic spillover effects from the economic regions to the local, national, and international levels. These effects can also be identified in the socio-economic dimension. Economic management alone is equivalent to a multi-sectoral framework that includes an ecosystem of various industries, diverse businesses, and different departments and organizations. This framework has a high semantic and functional affinity with risk management. From an institutional perspective, hybrid economic zones foster an efficient and diverse ecosystem of services and organizations.

Property Management: Property management involves a comprehensive process that encompasses various properties and rights with economic value, generating income from them. The intersection of asset management, economic resilience, and HEZ is encapsulated in a variety of investments, resources, products, values, and interests. Diversity represents the primary characteristic of hybrid economic zones, showcasing various assets. Proper management of HEZ assets can ensure economic resilience.

Control and Supervision Management: The control and supervision process is one of the most crucial processes proposed to measure whether actions are being followed according to the goals. It matches what should be with what is, what is desired with what is, and what is predicted with what is performed, providing a clear picture. Management, control, and supervision are mandatory at all stages of HEZ. They are effective factors in economic resilience that ensure its preservation.

2.5.2 Characteristics of Goods and Services

Service management is the backbone and a vital pillar in hybrid economic areas. The intersection of service management, economic resilience, and HEZ is summarized in the necessary diversity in services and goods and different activity nodes, the efficient and effective way of service management, how to play the operational role of services, and how to produce added value. This is reflected at other levels, especially in socio-economic needs. Basic services and secondary services should be implemented and managed in harmony and parallel with each other. The same perspective should be applied to both, considering that they are both essential in meeting the needs of communities. While ensuring the stability of core services and basic operations, secondary services should also be supported. Service management and asset management share significant similarities in terms of benefits and products, both contributing to enhancing economic resilience in hybrid economic environments.

2.5.3 Features of Society

Civil societies in hybrid economic zones play a crucial role, particularly in the local economy, where they are recognized as key stakeholders. On the other hand, they are engaged in various jobs as part of the labor force, and the surplus of knowledge and skills spreads among the local communities, benefiting both the national and international economy.

2.5.4 Characteristics of Activists

Actors and activists form the core of hybrid economic regions, which are defined by the influence of their policies, interests, goals, and strategies. The success of interventions in the fields of planning, policy-making, implementation, monitoring, and evaluation depends on the collaboration of all stakeholders. These actors include public and private entities responsible for macro policies in HEZs and can alter the current environment. In this area, the public sector is responsible for setting and formulating strategies aimed at the main services, including the oil and gas industry and related industries. The centralized system should remain in this sector. Secondary services, such as those related to tourism industries and local productions, can be implemented under the supervision of the private sector, with a combination of public–private management applied.

2.6 Concluding Remarks

Iran's public discourse, excluding the special economic zones of South Pars in Bushehr province, indicated that although these zones were established with the aim of economic profitability and under distinct commercial laws. The analysis revealed that Special Economic Zones (SEZs) have been established with single-function characteristics, a management system confined to the public sector, and a focus solely on the development of upstream, mixed, and downstream industries in the oil and gas sector, primarily aimed at exporting oil and gas to foreign nations.

In recent years, the intensification of economic instabilities and shocks on the economies of cities and countries has had a negative impact, leading to a shift towards prioritizing economic resilience in policymaking, particularly in special economic zones. South Pars Special Economic Zone's one-sided dependence on international markets and its integration with global economies make it vulnerable to the economic conditions of export destination countries, leading to exposure to shocks, crises, and damages. This vulnerability has historically resulted in production halts during the Iran-Iraq war. The lack of diversity acts as a hindrance to economic growth in special economic zones. This, coupled with weak institutional frameworks and financial obstacles, leads to stagnation in the country's economy. For this reason, it is important to pay attention to economic resilience and its drivers. Hybrid economic zones are defined as catalysts for economic resilience, providing robust support and ensuring it through their influence on the local, national, and international economies. They establish the foundation for collaborative investments.

One of the important pillars of hybrid economic zones is the design of the institutional framework that outlines the measures and factors associated with the management of HEZs. In the current research, the institutional framework of hybrid economic zones is examined, focusing on four dimensions of institutional characteristics (risk management, economic management, transformation management, public policy management, control and supervision management, asset management, and government organization design management); The study expands on the characteristics of goods and services (service management), the characteristics of actors (public–private actors), and the characteristics of society (civil societies and local funds). It shows that these four dimensions are interrelated, creating an efficient institutional framework for the management of hybrid economic regions. This framework helps South Pars become an economic and resilient region, capable of maintaining its resistance through a variety of services in times of crisis. The institutional framework presented provides the context for the transition from SEZ to HEZ in South Pars. It illustrates an economic region characterized by diversity, multifunctional activity nodes, private-state management, and the involvement of civil societies in the local economy, as well as cooperation at national, local, and international levels.

Acknowledgements In this section, we would like to express our gratitude to Professor Cheshmehzangi and all the editors for providing us with this opportunity.

References

1. Aggarwal A (2022) Special economic zones in the Indonesia-Malaysia-Thailand growth triangle. Asian Development Bank
2. Aitani K, Arima T Urban catalyst: regeneration of downtown area through community development, pp 223–233
3. Betts T, Buzzanell PM (2022) Enacting economic resilience: a synthesis of economic and communication frameworks. J Risk Finan Manage 15(4):178
4. Crane B, Albrecht C, Duffin KM, Albrecht C (2018) China's special economic zones: an analysis of policy to reduce regional disparities. Reg Stud Reg Sci 5(1):98–107
5. Dawei W (2021) SEZs and economic resilience: lessons from Covid Crisis. UNCTAD
6. Delimatsis P (2021) Financial services trade in special economic zones. J Int Econ Law 24(2):277–297
7. Dimitropoulos G (2021) International commercial courts in the 'modern law of nature': adjudicatory unilateralism in special economic zones. J Int Econ Law 24(2):361–379
8. Hallegatte S (2014) Economic resilience: definition and measurement. World Bank Policy Research Working Paper (6852)
9. Hu X, Yang C (2019) Institutional change and divergent economic resilience: path development of two resource-depleted cities in China. Urban Studies 56(16):3466–3485
10. Joseph T, Awolaja O, Ajibola O (2024) Measuring Sub-Saharan Africa economic resilience to external shocks: the role of adaptive policy space. Appl J Econ Manage Social Sci 5(1):1–28
11. Kahsai M, Yu J, Middleton M, Schaeffer PV, Jackson R (2015) A framework for measuring county economic resilience. Regional Research Institute Working Papers. 22, pp 1–21
12. Kalogiannidis S, Kalfas D, Chatzitheodoridis F, Lekkas E (2023) Role of governance in developing disaster resiliency and its impact on economic sustainability. J Risk Finan Manage 16(3):151
13. Lee S, Wang S (2023) Impacts of political fragmentation on inclusive economic resilience: examining American metropolitan areas after the great recession. Urban Stud 60(1):26–45
14. Mulenga R (2024) Navigating global economic volatilities: towards economic resilience. J Econ Finan Manage 3(1):295–308
15. Ostrom E (2011) Background on the institutional analysis and development framework. Policy Stud J 39(1):7–27
16. Putri I, Triansyah FA, Dinantika AR, Gunawan A (2023) Agility analysis to review economic resilience in micro, small and medium enterprises. Asian J Manage Entrepr Social Sci 3(01):579–597
17. Rose AZ (2009) Economic resilience to disasters. Final Report to Community and Regional Resilience Institute, pp 1–40
18. Sosnovskikh S (2017) Industrial clusters in Russia: the development of special economic zones and industrial parks. Russian J Econ 3(2):174–199
19. Trippl M, Fastenrath S, Isaksen A (2023) Rethinking regional economic resilience: preconditions and processes shaping transformative resilience. Eur Urban Reg Stud 09697764231172326
20. Umiński S, Nazarczuk JM, Borowicz A (2023) The role of foreign-owned entities in building economic resilience in times of crisis: the case of European digital and technologically-intensive firms during the Covid-19 pandemic. Equilibr Q J Econ Econ Policy 18(3):751–777
21. Zeng DZ (2021) The past, present, and future of special economic zones and their impact. J Int Econ Law 24(2):259–275

Chapter 3
Temporary Appropriation As a Catalyst for Diversity in Urban Development

Jose Antonio Lara-Hernandez

Abstract This study explores the concept of temporary appropriation (TA) and its profound impact on enhancing diversity within urban environments, contributing to the multidisciplinary dialogue of our cities. Temporary appropriation, the process by which individuals or communities repurpose urban spaces for alternative uses, embodies the dynamic interplay between diverse social groups and the built environment. This phenomenon not only reflects the rich tapestry of urban everyday life but also acts as a critical driver of social cohesion and urban resilience. Drawing from urban design, social sciences, and sustainability studies, this research delves into case studies spanning global contexts such as Venice, Mexico City, Hong Kong, and Auckland, emphasizing the role of temporary appropriation in fostering inclusivity, stimulating economic activities, and enhancing community resilience. By focusing on how diverse groups engage with and transform urban spaces for cultural expressions, community gatherings, and grassroots economic initiatives, the study highlights temporary appropriation as an essential mechanism for integrating diversity into the fabric of urban development. Moreover, this work argues that temporary appropriation can serve as a powerful tool for urban planners and policymakers to recognize and harness the potential of diversity. It encourages the adoption of flexible urban design strategies that accommodate the evolving needs and creative expressions of diverse urban populations, thereby promoting a more inclusive, sustainable, and resilient urban future.

Keywords Temporary appropriation · Informal practices · Diversity · Street network analysis

J. A. Lara-Hernandez (✉)
Future Environments, Auckland University of Technology, Auckland, New Zealand
e-mail: Antonio.lara-hernandez@aut.ac.nz

G. Siew et al. (eds.), *Diversity as Catalyst: Economic Growth and Urban Resilience in Global Cityscapes*, Urban Sustainability, https://doi.org/10.1007/978-981-97-8776-0_3

3.1 Introduction

Temporary appropriation (TA) is a worldwide phenomenon that refers to the creative and sometimes unexpected ways in which urban space is used. Lara-Hernandez and Melis [1], and Melis et al. ([2], Chapter 8) conceptualise TA as the act in which an individual or group temporarily appropriates public spaces for activities other than the original purpose for which they were designed. Lara-Hernandez and Melis [3] categorise such activities in three main groups: (1) work and economic such as selling products or services; (2) leisure and sports such as playing music or football; and (3) religious or sacralisation such as parades, praying and wordship. TA is a bottom up phenomena, having its roots in the so called "informal". Unlike movements like place-making, TA is unplanned, spontaneous, and has a strong cultural component which enhances the bond between people and places while forming urban identity. The study of TA is relevant to understand urban dynamics at the scale of public spaces because it is an indicator of urban diversity, social cohesion, and urban resilience [4]. The following section define these key concepts and details the conceptual framework.

3.2 Conceptual Framework

This section defines and discusses the theoretical background of key concepts utilized in this chapter, such as temporary appropriation, urban diversity, social cohesion, and urban resilience, and how all of them are strongly linked to the urban fabric. The term appropriation was first introduced in the field of environmental psychology by Korosec-Serfaty in 1976 during the Proceedings of the Strasbourg conference. It was described as a temporary phenomenon that implies a dynamic process of interaction between the individual and its surroundings [5]. In the same vein Graumann [6] supported this argument stating that appropriation of the environment (either natural or human made) is an inborn socio-spatial need of the individual. Yory [7], corroborates this argument by stating that we as human beings are naturally engaged with the construction (i.e. appropriation) of our environment. Similarly, sociologists such as Torres [8], Pol-Urrutia [9], Buffalo [10], and Pol and Iñiguez [11] have theorized that the appropriation allows citizens to produce urban space rather than just to use the already produced urban space.

Regarding the urban environment, Lara-Hernandez and Melis [1] noted that appropriation has a temporal dimension when it comes to the use of public spaces. Furthermore, TA can be interpreted as the interaction between users and the built-environment expressed through specified activities occurring in public spaces. According to Lara-Hernandez [4], these activities contribute to the formation of the social urban landscape [12] and are strongly related to the urban fabric [13]. Furthermore, Lara-Hernandez [14] states that the concept of TA is strongly linked to inclusivity, because different people use public spaces in a variety of ways. Consequently, the analysis of TA is an individual, social, and spatial need that must be taken in consideration when it comes to urban studies.

Although there is not a precise definition of the term, urban diversity encapsulates the variety of social, cultural, economic, and environmental attributes and experiences present within an urban context. Jacobs [15] was one of the first authors who stated diversity within the urban environment as an essential element of livable communities. In this context, she introduced the urban fabric as a decisive element fostering diversity, identifying four main characteristics such as primary functions (economic activities), short blocks (human scale), old buildings, and density. Even today, her ideas have influenced the urban discourse and practice under the banner of mixed use [16] and the so called New Urbanism (see Duany et al. [17]). Urban diversity is considered a critical asset that contributes to the vibrancy and resilience of cities, fostering innovation and adaptability. Similarly, the concept of urban diversity is deeply intertwined with the urban fabric [18, 19]. For instance, an heterogeneous urban fabric is associated to a more resilient urban environment [20].

Previous studies have focused on the diversity of the built environment related to the economic activities [21–23], the diversity of its population in terms of ethnicity, culture, and socio-economic status [24, 25], and the ecological diversity within the socio-urban environments [26]. In terms of scale, most existing research on diversity has focused on the city, regional, national, and international levels (e.g., Duranton and Puga [27]). Recently, Yoshimura et al. [28] focused on the relationship between urban morphology and diversity on different spatial scales, including the neighborhood, while quantifying the concept of city diversity. However, few studies have focused on the study of diversity of users at the scale of the public space [29, 30] and even less in relation to sense of community.

Sense of community refers to the strength of relationships and the sense of solidarity among members of a community [31–33]. Where member feel attached to the group. It involves the extent to which citizens trust each other, share common values, and engage in collective activities. A cohesive society can effectively cooperate to achieve mutual benefits, overcome challenges, and ensure that all members are included in social, economic, and political life. Strong sense of community is often seen as essential for sustainable urban development and resilience [34]. Sense of community is also strongly associated to the urban fabric. For instance, studies shows that people who live in buildings or city areas that have spaces allowing social encounters tend to have stronger social bonds among them [35–37].

Urban resilience is a term utilised in different fields such as ecology, urban studies, architecture, psychology, sociology and even technology. The term has become an everyday use word. From local to international level urban resilience has become a priority, including strategies for climate change adaptation. For instance, it has been included in the Agenda 2030 [38]. Urban resilience refers to the capacity of urban systems, communities, and individuals to survive, adapt, and grow in the face of physical, social, and economic challenges [39–41]. It involves not only resistance to shocks and stresses but also the ability to bounce back and positively adapt to change. Resilience thinking in urban contexts emphasizes the importance of flexibility, learning, and adaptability, integrating these principles into urban planning and development practices [42, 43].

3.3 Theoretical Background Linking the Concepts

The theoretical background linking these concepts emerges from interdisciplinary research across urban planning [44], social sciences [45], and ecological studies [46]. Several authors such as Gevorkyan et al. [47], Liang [48] claim the social realm is closely related to the concept of urban resilience. Strong social bonds among community members is commonly associated with more organised responses when facing crises [45, 49, 50]. For instance, Lara-Hernandez [42] demonstrates how the human scale planning in conjunction with strong sense of community in Venice contributed to the emergence of TA during last COVID-19 pandemic. TA diversity is associated to more inclusive public spaces where users' diversity is enhanced [51]. Moreover, when public space networks are present in the urban fabric promote urban diversity.

The intersection of temporary appropriation and urban diversity is particularly prominent in how diverse urban populations utilize and transform urban spaces in innovative ways that reflect and cater to their unique cultural and social needs [52]. This utilization acts as a driver for social cohesion by providing a platform for community engagement and interaction, fostering relationships among different social groups [34].

From a resilience perspective, diverse and cohesive communities are better equipped to respond to and recover from urban challenges such as economic downturns [53], social conflicts, and environmental crises [54]. The adaptive use of urban spaces through temporary appropriation can enhance urban resilience by creating flexible environments that can be quickly transformed to meet changing community needs and by promoting a sense of community ownership and stewardship over urban spaces [2].

In essence, temporary appropriation, by leveraging urban diversity, enhances social cohesion, which in turn strengthens urban resilience. This linkage underscores the importance of considering social and cultural dimensions in urban planning and policy-making to harness the potential of urban diversity as a resilience factor.

3.4 Methodology

The multi-dimensional nature of TA requires the use of diverse research methodologies, including observational techniques, recognized as pivotal in urban landscape analyses. Observational methods have long been favoured for studying dynamic urban scenarios [55–57], particularly due to their sensitivity to changes in land use and socioeconomic contexts [58–60]. We opted for a comparative case study approach, which is effective in understanding how culture and the use of the public space can influence TA.

In this research, TA activities were recorded on-site using an "activity mapping" technique, which assesses street-level interactions concerning land uses and physical elements [29, 61]. This approach is especially valuable for revealing the 'temporal city', highlighting the ongoing interactions between users and the urban landscape ([4], Chapter 3). This method, akin to that described by Groat and Wang [62], takes an ethnographic stance through participant observation.

Additionally, we incorporated a quantitative analysis of the street network, inspired by Geoff Boeing's methodological framework [63]. Utilizing street network analysis not only provides precise measurements of urban morphology but also allows for a more robust comparison of spatial patterns across different cities. This methodological approach enhances our ability to pinpoint how variations in street network design can impact both the intensity and type of temporary appropriations. Moreover, by quantifying complex network characteristics, this method offers clearer insights into how urban form influences social interactions and economic activities, which are critical to enhance urban liveability and sustainability. Thus, this analysis was conducted to understand the spatial configuration and connectivity of the urban areas under study. The following steps outline the integration of this street network analysis into our methodology:

1. **Data Collection**

    - The OSMnx library was used in Python to download street network data for each study area: Venice, Mexico City, Hong Kong, and Auckland CBD. The graph_from_place function was employed to retrieve the road networks, with the parameter network_type = 'drive' to capture vehicle roads.

2. **Network Analysis**

    - **Graph Construction and Simplification**: The street network data was processed to create a graph representation, simplifying it to retain only the primary roadways and intersections relevant to our study. This was achieved using the simplify_graph function in OSMnx.
    - **Projection and Standardization**: To facilitate spatial analysis, we projected the graph to a common coordinate system using UTM (Universal Transverse Mercator) zones appropriate for each city. The project_graph function in OSMnx was used to standardize the spatial reference.
    - **Topological Metrics**: We calculated key topological metrics, including node degree distribution, average path length, and network centrality measures (Betweenness Centrality, Closeness Centrality, and PageRank). These metrics provided insights into the structural properties and connectivity of the street networks.

3. **Spatial Visualization**
   - The street networks was visualised using plot_graph from OSMnx, enabling a comparative visualization of network densities, node centralities, and connectivity patterns across the study areas. This visualization helped identify spatial relationships and network characteristics pertinent to TA activities.
4. **Integration with Activity Mapping**
   - The results from the street network analysis were integrated with the activity mapping data to examine how the spatial configuration of streets influences TA activities. We used GIS software to overlay the activity maps onto the network graphs, highlighting areas with high interaction density and connectivity.
5. **Comparative Analysis**
   - The street network metrics and TA activity patterns were compared across the four cities, looking for correlations and distinctions in network topology and urban interaction dynamics. This comparative analysis aimed to uncover how different urban forms and network structures facilitate or hinder TA.

By incorporating this street network analysis, our study offers a comprehensive understanding of how spatial configuration influences the dynamics of TA in various urban settings. This methodological integration enhances our ability to draw nuanced conclusions about the role of urban design in fostering diversity and resilience through temporary appropriation.

The following section illustrates each of the case studies in terms of TA and street network analysis.

3.5 Global Case Studies Analysis

Given the cultural nuances of TA already emphasised, choosing the right cities is crucial to deepening our understanding of TA as a diversity driver. DeLanda [64] suggests that certain cities can be described as "high-coded cities" due to their distinct regional cultures and clear identities. Cities with a global stature exhibit a complex geographical tapestry [21], characterized by vibrant interactions among cultural, social, and physical dimensions [65]. This interplay of different urban layers, including social, legal, and cultural elements, characterises a unique urban space [66]. Brenner [67] further elaborates that social interactions are stratified in a "vertical" structure, interwoven through a hierarchical framework of territorial layers from global to supranational, national, regional, metropolitan, urban, local, down to the individual level. Therefore, we have chosen a city in each of the continents from different cultures, climate, scale and urban density such as Venice, Mexico City Centre, Hong-Kong, and Auckland to demonstrate this concept, as it exemplifies a highly coded city, serving as a practical illustration for one of the chapter's key arguments.

3.5.1 Venice

The city of Venice is a particular example because it has two built environment characteristics that are strongly related to TA such as a human scale and a double fractal urban fabric (see **Error! Reference source not found.**). Here the human scale is referred to the built-environment's configuration that allows to form the "space of contact" [68], in which social interaction occurs in the urban sphere. Narrow and variable streets' dimension, building's height less than 16 m, organic urban planning, rich and appealing streetscape, and mix-use building functions are common attributes related to the former. While the fractal urban fabric refers to organic forms with rich detail at multiple scales [69], having the water on the one hand and the built-environment on the other [34, 70].

This intricated urban fabric in conjunction with the cultural landscape (see [34]), and a completely pedestrian street use lays the way to a fertile environment, allowing TA to emerge in peculiar ways. Figure 3.1 shows the street network graph of Venice, having a remarkable presence of nodes which functions as space of contact. Figure 3.2 shows a carpenter to cutting a timber piece to fix a door in the street. By doing so the street temporarily also became the carpenter's outdoor workshop. Figure 3.3 illustrates diverse TA activities taking place under the *Ponte della Constituzione* (also known by the locals as *Ponte di Calatrava*) such as children playing by using the sloop as a slide, a couple having coffee, youngsters loitering and a woman utilising the city as a stage for taking photos of her child. Lastly, Fig. 3.4 illustrates once a normal scene in every city, children freely playing in the street.

3.5.2 Mexico City

Mexico City is one of the most populated metropolises in North America, having more than 23 millions of inhabitants. Mexico City Centre is one of the most emblematic and dynamic area in terms of cultural heritage either tangible or intangible. It is the biggest and the oldest colonial historic centre in America with 9.1 km^2 and founded in 1521. The urban planning pattern used in Mexico was applied to most of the Spanish colonies in America [71]. The urban fabric of Mexico City Centre is the syncretism between Tenochtitlan, the capital city of the Aztec Empire, and the master planning inspired by the Roman Grid. Thus, regarding human scale it is similar to Venice, sharing many of the attributes already listed. Figure 3.5 illustrates the street network graph of Mexico City Centre, showing a more even presence of nodes in relation to the edges.

Fig. 3.1 The graph of a Venice's street network. Circles are nodes and lines are edges. *Source* Author

Fig. 3.2 Carpenter improvising a workshop in the street. *Source* Author

Fig. 3.3 Diverse TA activities under Calatrava Bridge. *Source* Author

Fig. 3.4 Kids playing in San Giacomo del Oreo. *Source*: Author

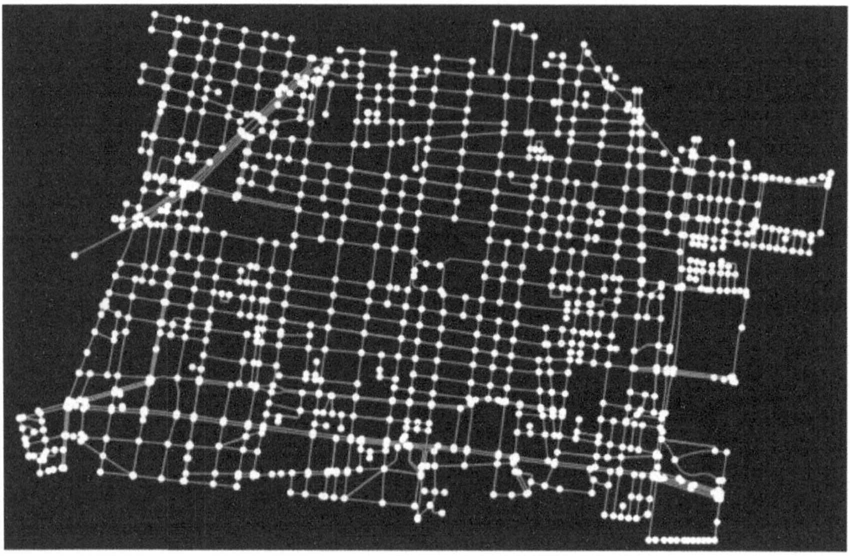

Fig. 3.5 The graph of a Mexico City street network. *Source* Author

This urban context in terms of urban fabric and cultural landscape already described elsewhere ([4], Chapter 4) allows TA to emerge in a variety of ways (Fig. 3.5). For instance, Fig. 3.6 illustrates a group of men utilizing the street for playing fronton in Mexico City Centre. The difference between the level of the sidewalk and the street allows to have a "walled court" that is ideal for playing. Similarly, Fig. 3.7 shows how a section street is temporarily transformed in an outdoor hairdresser salon, in which the bollard as an urban furniture became the "client seat". Figure 3.8 illustrates a mariachi band playing in the street as an example of artistic activities occurring on the street. Therefore, demonstrating the versatility, creativity, and flexibility of TA emergence in the urban realm.

3.5.3 Hong Kong

The city of Hong Kong is the forth most densely populated city worldwide, having 7065 inhabitants per km^2 of land area [72]. The city is compact, having vertically zoned spaces connected through vertical and horizontal transitional spaces like lifts, escalators, bridges, underpasses, and stairs result in dynamic patterns of horizontal and vertical pedestrian movement [73]. This pedestrian experience could be similar to a few cities in the world such as Singapore and central areas of Montreal. Although not at the human scale, this attributes in conjunction with mix-use land building functions, and the high-capacity mass transit railway system favors a very active and

Fig. 3.6 People playing fronton on Mexico City Centre. *Source* Melis et al. ([2], Chapter 12)

lively urban landscape. Figure 3.9 illustrates the street network graph of Hong-Kong, which again illustrates a remarkable existence of nodes.

For instance, Fig. 3.10 illustrates an altar in the street. A foreigner is not normal to recognize the cues in the urban landscape, while locals identify that part of the street as a sacred place, thus, showing the proper respect when passing by. Here place is understood as what Relph [74] described as a space embedded with meaning, already discussed elsewhere [13, 75, 76]. Figure 3.11 illustrates a woman selling clothes utilizing the sidewalk and the scaffoldings as an effective display framing and promoting her merchandise. Figure 3.12 illustrates a man exercising in the street as it he was in his private gym.

3.5.4 Auckland

The Central Business District of Auckland, New Zealand, is one of the most extensive historical areas of the city, with a total area of 4.33 km^2 [77] and a population of 22,804 inhabitants [78]. The CBD was the first land bought from Ngāti Whātua by the Crown in 1840. It covers 433 hectares in a triangular area bounded by the

Fig. 3.7 Street hair studio. *Source* Melis et al. [2]

Waitematā Harbour and the inner suburbs of Ponsonby, Newton and Parnell. The buildings with heritage value in Auckland's CBD were almost entirely created after the 1840s by European immigrants, and the street layout of Auckland's CBD would have been influenced by the regulations and models that were laid out by the Colonial Secretary's Office (1829) in Australia ([79], p. 90). However, the current plot pattern is the result of a combination of the amalgamation, subdivision and truncation of plots [80]. Thus, such urban fabric is not designed to the human scale.

During the 1990s, education became the CBD's biggest export earner. The University of Auckland and Auckland University of Technology, as well as new English language schools, catered for large numbers of Asian students [78]. The inner city was revitalised by these incomers, and others drawn to apartment life in the city. Moreover, Auckland's CBD is one of the best examples of a multicultural area within the city [81, 82] and this defines its particular identity. However, despite such multicultural environment it has been reported that the sense of community among its residents is weak [83]. Figure 3.13 illustrates Auckland CBD's street network graph. The presence of nodes is scarce compared to cities like Venice, Hong-Kong or Mexico City, thus reducing the possibilities of space of contact. Figures 3.14 and 3.15 show some examples of TA in Auckland CBD.

Fig. 3.8 Mariachi band playing. *Source* Author

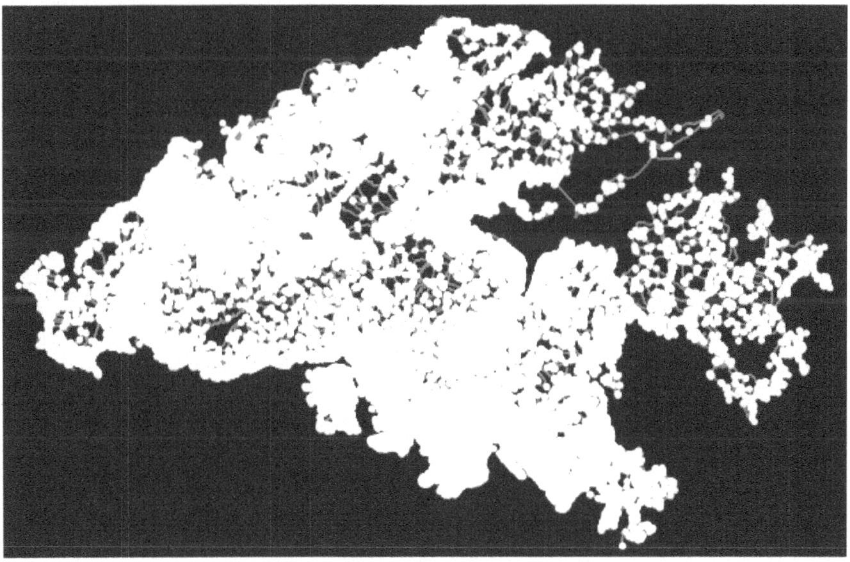

Fig. 3.9 The graph of a Hong-Kong street network. Circles are nodes and lines are edges. *Source* Author

Fig. 3.10 Altar in Hong Kong. *Source* Author

3.6 Comparative Analysis

Prior studies that have noted the importance of the use of public spaces through its temporary appropriation for creating lively communities. Urban street networks are vital for understanding the spatial layout, connectivity, and accessibility of cities. Geoff Boeing's methodologies provide robust frameworks to analyze these networks, focusing on metrics such as street density, centrality, and network efficiency. This analysis applies these methods to four distinct cities: Venice, Mexico City, Hong Kong, and Auckland. Each city's street network reveals unique characteristics influenced by historical development, geography, and urban planning. Table 3.1 illustrates the street network values of each of the selected cities.

Venice, offers a distinct example of TA shaped by its unique urban form and historical context. The city's intricate canal system and narrow pathways facilitate intimate and localized forms of appropriation, such as small floating markets or impromptu musical performances on bridges. These activities are deeply embedded within the fabric of the city, enhancing the experiential quality of its spaces and fostering a strong sense of place among residents. Venice's approach to TA contributes to its resilience

Fig. 3.11 Woman selling clothes. *Source* Author

by utilizing its architectural uniqueness to maintain a robust cultural identity and tourist economy despite environmental challenges.

Venice has an intricate network of canals and narrow streets, which is characterized by a dense network with a high number of intersections (15,989) relative to its node count (20,867), indicating a complex urban grid that facilitates pedestrian movement. The average number of streets per node (2.69) is lower than the other cities, emphasizing Venice's numerous dead-ends or alleys typical of a historic pedestrian-focused city. The street lengths are shorter on average (51.26 m), which aligns with the compact urban form of Venice.

This city's layout is a product of its historical development as a maritime republic.

1. **Network Density and Connectivity**

   - **Street Density**: Venice has an extremely high street density, characterized by its narrow pathways and limited vehicular access. The average street width is narrow, facilitating pedestrian movement and enhancing spatial density.
   - **Connectivity**: The network is highly connected but challenging for vehicles. The lack of cars and the prominence of foot traffic contribute to a high degree of local connectivity. Venice's street network is a web of alleys and small squares, promoting pedestrian flow and social interaction.

Fig. 3.12 Man practicing Tai-chi. *Source* Author

2. **Network Centrality**
   - Centrality metrics such as betweenness centrality and closeness centrality are high in Venice's historic core, reflecting its status as a central node in the city's social and economic life.
3. **Spatial and Historical Considerations**
   - Venice's layout, shaped by its island geography and historical needs for defense and trade, results in a highly irregular grid that prioritizes human scale and environmental integration over vehicular flow.

In Mexico City Centre, temporary appropriation reflects the city's rich historical layers and vibrant public life. The city centre, marked by its urban/architectural syncretism and human scale, serves as a platform for various forms of TA that echo its deep cultural heritage. Here, the streets transform into markets where local artisans and vendors exhibit traditional crafts and foods, directly linking the economy with cultural expression. This integration of the so called informal economic activities within the urban core not only fosters a lively street culture but also strengthens social bonds among the community, demonstrating the city's dynamic approach to urban diversity and resilience [4].

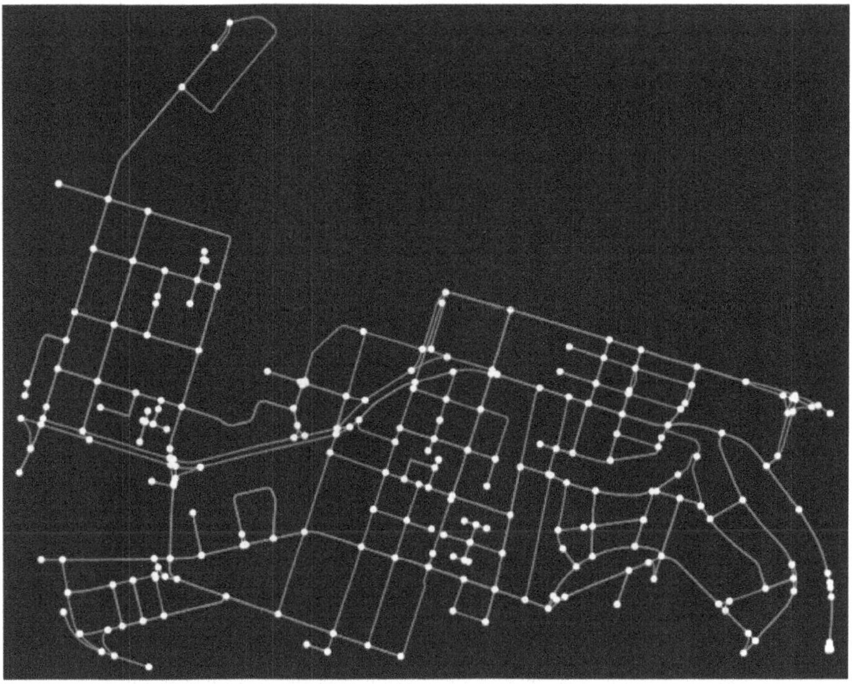

Fig. 3.13 The graph of a Auckland CBD street network

Mexico City Centre presents a different configuration with fewer nodes (1151) but a higher average connectivity per node (3.42 streets per node), suggesting a grid-like network typical of colonial city planning. The streets are longer on average (97.21 m), indicating larger blocks typical of Spanish colonial cities. The network's total edge length (211,938 m) is considerably less than Venice's, reflecting its smaller scale of detailed network analysis.

These data points illustrate the varying approaches to urban planning across different contexts—Venice's historical pedestrian-focused pathways, Mexico City's colonial heritage with broader avenues, Hong Kong's hyper-dense multi-level crossings, and Auckland's modern, less congested streets. Each city's network characteristics profoundly influence the urban experience, affecting everything from daily commutes to the capacity for public gatherings and the overall resilience of the urban fabric.

1. **Network Density and Connectivity**

   - Street Density: Mexico City's network is less dense than Venice's but still exhibits significant complexity, especially in the historic centre. The city's grid system, combined with radial avenues, creates a diverse urban fabric.

Fig. 3.14 Man playing the plute in Queen St. Auckland, New Zealand. *Source* Author

- Connectivity: The connectivity is high, with a mix of arterial roads and smaller streets. The central area, particularly around the Zócalo, showcases high connectivity, while the outskirts have lower connectivity due to urban sprawl.

2. **Network Centrality**

- The central areas exhibit high betweenness and closeness centrality, indicating significant interconnectivity. Major avenues like *Paseo de la Reforma* and the network around the historic centre are central to the city's flow.

3. **Spatial and Historical Considerations**

- The city's development, influenced by both pre-Hispanic and colonial planning, has resulted in a network that balances historical integrity with modern infrastructure demands. Urban sprawl has impacted the network's connectivity in peripheral zones.

Hong Kong, with the largest node count (114,400) and the highest intersection density (91,049 km), displays an extremely dense and complex urban grid, driven

Fig. 3.15 Monk predicating. *Source* Author

Table 3.1 Street network values

City	Auckland CBD	Hong Kong	Mexico City	Venice
Node count (n)	248	114,400	1151	20,867
Edge count (m)	550	315,956	2221	56,100
Average node degree (k_avg)	4.44	5.52	3.86	5.37
Total edge length	45,726.435 m	19,543,962.23 m	211,938.556 m	2,875,763.768 m
Average edge length	83.14 m	61.86 m	95.42 m	51.26 m
Average streets per node	2.97	2.76	3.42	2.69
Intersection count	219	91,049	1089	15,989
Total street length	28,892.186 m	9,771,981.115 m	182,362.705 m	1,437,881.884 m
Average street length	82.08 m	61.86 m	97.21 m	51.26 m
Circuity average	1.04	1.13	1.01	1.08
Self-loop proportion	0.0	0.0026	0.0	0.0033
Node density (per km^2)	248	114,400	1151	20,867
Intersection density (per km^2)	219	91,049	1089	15,989
Edge density (per km^2)	45,726.435 m	19,543,962.23 m	211,938.556 m	2,875,763.768 m
Street density (per km^2)	28,892.186 m	9,771,981.115 m	182,362.705 m	1,437,881.884 m

by its status as a major global city with significant vertical and horizontal expansion. The streets per node average (2.76) is similar to Venice's, but the network handles a much larger volume of traffic and pedestrian flow, evidenced by the vast total edge length (19,543,962 m).It presents a contrast to both cities with its high-density urban environment and vertical urbanism. TA occurs in the interstitial spaces of this high-rise city, where ground level, rooftops, and even air bridges become sites for social and economic interactions. For instance, open areas beneath apartment complexes often serve as informal playgrounds or social gathering spots, adapting to the dense urban form. The city's use of vertical space for TA, such as rooftop gardens and outdoor fitness classes, showcases an innovative approach to urban design that accommodates the diverse needs of its dense population. This adaptability not only alleviates the pressures of urban living but also strengthens community resilience by creatively using every available space.

Hong Kong's street network is a complex blend of high-density urban areas and more open spaces, shaped by its topography and colonial history.

1. **Network Density and Connectivity**
   - **Street Density**: Hong Kong's network is highly dense, particularly in urban areas like Central and Kowloon. The streets are often narrow, with high-rise buildings lining the roads.
   - **Connectivity**: The network is extremely well-connected, with a high degree of redundancy. The presence of multiple levels of roads, including elevated highways and pedestrian overpasses, enhances connectivity.
2. **Network Centrality**
   - Centrality metrics highlight the dense clustering of nodes in urban areas. The network around Central and Tsim Sha Tsui is central, with high betweenness and closeness centrality.
3. **Spatial and Historical Considerations**
   - Hong Kong's network is heavily influenced by its hilly terrain and colonial layout. The adaptation to topographical challenges and the integration of modern infrastructural developments are key features.

Auckland CBD has the smallest network analyzed, with only 248 nodes and an average street length of 83.14 m, which suggests a newer, less dense urban layout. Its average streets per node (2.97) is higher than Venice but lower than Hong Kong, indicative of a less complex network that likely supports a mix of pedestrian and vehicular traffic. This case study is relevant because the urban fabric is not up to the human scale because it has a recent history. The phenomenon of temporary appropriation is evident in the spatial dynamics shaped significantly by educational and economic developments. The presence of major educational institutions, notably the University of Auckland and Auckland University of Technology, has introduced a substantial population of international students, catalysing shifts in local demographics and urban usage patterns. This academic influx has led to the emergence of multifunctional spaces that accommodate both transient and semi-permanent activities, ranging from seasonal markets to cultural festivals. Such transformations in the urban fabric highlight the adaptability of public spaces to meet diverse community needs.

Moreover, the CBD's evolution is also marked by a strategic engagement with urban planning initiatives aimed at enhancing liveability and accessibility. The redevelopment efforts, particularly around waterfront areas and major thoroughfares like Queen Street, have been designed to foster greater community interaction and economic activity, thereby facilitating a variety of temporary appropriations. These initiatives not only reflect the city's response to changing urban demands but also illustrate the broader implications of temporary appropriation in enhancing urban resilience and social cohesion. This case study of Auckland's CBD underscores the critical role of urban design and policy in shaping the contexts in which temporary appropriations occur, thereby contributing to a more dynamic and inclusive urban environment.

Auckland's street network reflects its development as a modern city with a relatively low-density layout compared to the other cities.

1. **Network Density and Connectivity**
   - **Street Density**: Auckland's network is less dense, with wider streets and significant open spaces. The CBD features a grid pattern, while the suburbs are more sprawling.
   - **Connectivity**: The connectivity is moderate, with main arterials and a grid layout in the CBD. However, suburban areas show lower connectivity, with significant car dependency.
2. **Network Centrality**
   - Centrality is moderate, with high connectivity in the CBD. The streets around Queen Street and the waterfront are central, but overall network centrality is less pronounced compared to the other cities.
3. **Spatial and Historical Considerations**
   - Auckland's network is shaped by its geography and planning policies, emphasizing car accessibility and suburban expansion. The city's development has been influenced by a focus on residential living and motor vehicle infrastructure.

Comparative Insights

The street network data for Venice, Mexico City, Hong Kong, and Auckland CBD reveals significant differences in urban connectivity and layout, reflecting each city's unique spatial and historical context.

1. **Street Density**: Venice has the highest street density, followed by Hong Kong, Mexico City, and Auckland. The dense networks in Venice and Hong Kong facilitate high pedestrian movement, while Mexico City's and Auckland's networks are more spread out.
2. **Connectivity**: Hong Kong exhibits the highest connectivity, with a complex network of roads and pathways. Venice, with its pedestrian-focused layout, also scores highly in local connectivity. Mexico City's central areas are highly connected, while Auckland's suburban areas are less so.
3. **Network Complexity**: Venice's irregular and narrow streets, Hong Kong's layered urban design, Mexico City's mix of historic and modern planning, and Auckland's grid pattern and suburban sprawl each reflect their unique historical and geographical contexts.

The street networks of Venice, Mexico City, Hong Kong, and Auckland illustrate the diverse ways cities develop and function. Venice's pedestrian-friendly, dense network stands in contrast to the sprawling, car-centric layout of Auckland. Mexico City and Hong Kong occupy a middle ground, each with its own blend of historical and modern influences. These differences underscore the importance of contextual factors in shaping urban environments and highlight the need for tailored urban planning strategies to enhance connectivity and sustainability in diverse cities.

3.7 Discussion

The case studies examined in this research show the multifaceted role of TA in enhancing urban diversity and resilience. Through a close analysis of cities like Venice, Mexico City, New York, Hong Kong, Nigeria, and Auckland, the study has highlighted how TA serves not only as a practical response to the immediate needs of urban populations but also as a strategic instrument in urban planning and policy formulation.

In cities like Venice and Mexico City, TA emerges from a rich historical context where the blending of cultural narratives with human scale urban fabric facilitates more diverse forms of TA. These appropriations are not merely acts of survival but are integral to maintaining cultural identity and heritage. For instance, Venice's adaptation of its intricate canal system for commercial and cultural activities underlines the resilience of its urban fabric against environmental and economic challenges.

Conversely, in a city like Hong Kong, where vertical urban dynamics dominate, TA reflects the necessity to utilize every available space, whether it be rooftops, edges, bridges, or underpasses. This form of TA highlights the city's response to its high-density challenges, showcasing an innovative integration of vertical and horizontal spaces to enhance social interaction and community bonding.

The research also points to a significant correlation between TA and urban resilience. In New York and Auckland, where urban spaces are continually reshaped by diverse demographic shifts and economic demands, TA acts as a catalyst for social cohesion. It provides a platform for diverse communities to express their cultural identities and economic ventures, thereby fostering a more inclusive urban ecosystem.

Furthermore, the analysis underscores the importance of policy frameworks that support the flexibility of urban spaces to adapt to the needs of their dynamic populations. Urban planning that incorporates the principles of TA can lead to the creation of cities that are not only physically adaptive but also socially and economically robust.

In conclusion, this study reaffirms that temporary appropriation is a pivotal element in the discourse of urban development. It challenges traditional perceptions of urban space usage and encourages a rethinking of how cities can be designed and governed to embrace the diversity and dynamism that define contemporary urban life. Moving forward, it is imperative that urban planners and policymakers consider the lessons of TA to ensure that cities remain resilient and responsive to the needs of their ever-evolving communities.

The case studies examined in this research show the multifaceted role of TA in enhancing urban diversity and resilience. Through a close analysis of cities like Venice, Mexico City, Hong Kong, Nigeria, and Auckland, the study has highlighted how TA serves not only as a practical response to the immediate needs of urban populations but also as a strategic instrument in urban planning and policy formulation.

3.8 Conclusion

This study has explored the multifaceted role of temporary appropriation (TA) in enhancing urban diversity and resilience. Through the analysis of diverse global contexts, including Venice, Mexico City, Hong Kong, and Auckland, it has become evident that TA acts as a dynamic interface between urban populations and their environments. These interactions not only cater to immediate community needs but also foster sustainable urban development by integrating diverse cultural, social, and economic activities into the urban fabric.

The case studies presented illustrate that TA can significantly influence urban planning and policy-making. In Venice, the unique adaptation of its historical and compact urban layout for pedestrian use highlights TA's role in preserving cultural heritage while supporting modern urban needs. Meanwhile, Mexico City demonstrates how TA can blend historical elements with contemporary demands to enhance both cultural vibrancy and economic vitality. In Hong Kong, the vertical integration and efficient use of limited urban space underscore the importance of innovative approaches to TA in high-density settings. Lastly, Auckland shows that even in less dense urban environments, TA can enhance liveability and connectivity, proving its relevance across different urban contexts.

Furthermore, the findings suggest that urban resilience is markedly enhanced by the inclusivity and flexibility facilitated by TA. Cities that incorporate TA into their urban design and policy frameworks are better equipped to respond to and evolve with the diverse needs of their inhabitants. This resilience is particularly pertinent in the face of global challenges such as urbanization, climate change, and social inequality.

In conclusion, temporary appropriation is not merely a byproduct of urban life but a crucial tool for promoting diversity and resilience. Urban planners and policymakers are encouraged to recognize the value of TA and to deliberately incorporate its principles into urban development strategies. By doing so, cities can become more adaptable, inclusive, and vibrant—qualities that are essential for the sustainable development of urban spaces in the twenty-first century.

References

1. Lara-Hernandez JA, Melis A (2018) Understanding the temporary appropriation in relationship to social sustainability. Sustain Cities Soc 39C:1–14. https://doi.org/10.1016/j.scs.2018.03.004
2. Melis A, Pievani T, Lara-Hernandez JA (2024) Architectural exaptation: when function follows form. Routledge
3. Lara-Hernandez JA, Melis A, Caputo S (2017) Understanding spatial configuration and temporary appropriation of the street in Latin American cities: the case of Mexico City Centre. In: Boudagh H, Versaci A, Trapani F, Migliore M, Clark N, Sotoca A (eds) Urban planning and architectural design for sustainable development. IEREK, Springer, pp 153–173
4. Lara-Hernandez JA (2019) Temporary appropriation: theory and practice of the street (Issue August 2019) [University of Portsmouth]. https://researchportal.port.ac.uk/portal/en/theses/temporary-appropriation(62a30252-cbb1-457a-b751-52b66176d8d7).html
5. Korosec-Serfaty P (1976) Appropriation of space. In: Proceedings of the Strasbourg conference. IAPC-3
6. Graumann CF (1976) The concept of appropriation (Aneignung) and modes of appropriation of space. In: Appropriation of space
7. Yory MC (2011) El Concepto de Topofilia entendido como Teoría del Lugar. Revista Barrio Taller, pp 1–17
8. Torres E (2009) Apropiación Versus Usos Del Espacio Público. https://www.mardelplata.gob.ar/documentos/policia/taller1culturayterritorioapuntes.pdf
9. Pol Urrútia E (2002) El modelo dual de la apropiación del espacio. Psicología y Medio Ambiente. Aspectos Psicosociales, Educativos y Metodológicos, pp 123–132
10. Búffalo L (2008) El uso del espacio público y la apropiación privada del espacio en la ciudad de Córdoba. Proyección 2(5)
11. Pol E, Iñiguez L (1996) La Apropiacion del Espacio. Monografías Psicosoaicles Ambientales 9
12. Lara-Hernandez JA, Chin JJ-Y (2022) Sense of community and spatial agency: key elements of resilient communities. In Carta M, Perbellini MR, Lara-Hernandez JA (eds) Resilient communities and the Peccioli charter: towards the possibility of an Italian charter for resilient communities. Springer International Publishing, pp 59–64. https://doi.org/10.1007/978-3-030-85847-6_7
13. Lara-Hernandez JA, Coulter CM, Melis A (2020) Temporary appropriation and urban informality: exploring the subtle distinction. Cities 99(102626):1–10. https://doi.org/10.1016/j.cities.2020.102626
14. Lara-Hernandez JA (2023) Temporary appropriation of public spaces as an indicator of social resilience. In: Melis A, Lehmann S (eds) Comunità Resilienti: Una Raccolta di Brevi Saggi Sulla Capacita' Adattiva delle Citta, vol 2. D Editore, pp 163–173. https://www.comunitaresilienti.com/

15. Jacobs J (1961) The death and life of great American cities. Random House
16. Bentley I, Alcock A, Murrain P, McGlynn S, Smith G (1985) Responsive environments. The Architectural Press Ltd.
17. Duany A, Plater-Zyberk E, Speck J (2000) Suburban nation. North Point Press, New York
18. Moulaert F, Demuynck H, Nussbaumer J (2004) Urban renaissance: from physical beautification to social empowerment. City 8(2):229–235. https://doi.org/10.1080/1360481042000242175
19. Rapoport A (1982) Urban design and human systems: on ways of relating buildings to urban fabric. In: Laconte P (ed) Human and energy factors in urban planning: a systems approach. Martinus Nijoff Publishers, pp 161–184. https://doi.org/10.1007/978-94-009-7651-1
20. Fort BL (2019) Heterogeneous urban fabrics. Typo-morphological diversity as enabler of urban resilience. In: The 2nd international European urbanisms seminar, Leuven. https://dial.uclouvain.be/pr/boreal/object/boreal%3A224100/datastream/PDF_01/view
21. Fyfe NR, Kenny JT (2005) The urban geography reader. In: The Routledge urban reader series. http://www.loc.gov/catdir/toc/ecip0421/2004018547.html%5Cn, http://www.loc.gov/catdir/enhancements/fy0650/2004018547-d.html
22. Glaeser E, Kolko J, Saiz A (2001) Consumer city. J Econ Geogr 1(1):27–50. https://doi.org/10.1093/jeg/1.1.27
23. Ribeiro PJG, Pena Jardim Gonçalves LA (2019) Urban resilience: a conceptual framework. Sustain Cities Soc 50(November 2018). https://doi.org/10.1016/j.scs.2019.101625
24. Burchardt M, Höhne S (2016) The infrastructures of diversity: materiality and culture in urban space. New Divers 17(2):1–3. https://pure.mpg.de/rest/items/item_2306983/component/file_2306981/content
25. Jones DE (2014) The definition of diversity: two views. A more inclusive definition. In: Managing multiculturalism and diversity in the library: principles and issues for administrators, vol 826, pp 5–15. https://doi.org/10.1080/01930826.1999.12113171
26. Cohen M, Baudoin R, Palibrk M, Persyn N, Rhein C (2012) Urban biodiversity and social inequalities in built-up cities: new evidences, next questions. The example of Paris, France. Landsc Urban Plan 106(3):277–287. https://doi.org/10.1016/j.landurbplan.2012.03.007
27. Duranton G, Puga D (2000) Diversity and specialisation in cities: why, where and when does it matter? Urban Stud 37(3):533–555
28. Yoshimura Y, Kumakoshi Y, Milardo S, Santi P, Arias JM, Koizumi H, Ratti C (2022) Revisiting Jane Jacobs: quantifying urban diversity. Environ Plann B Urban Anal City Sci 49(4):1228–1244. https://doi.org/10.1177/23998083211050935
29. Lara-Hernandez JA, Melis A, Coulter CM (2018) Using the street in Mexico City Centre: temporary appropriation of public space vs legislation governing street use. J Public Space 3(3):25–48. https://doi.org/10.32891/jps.v3i3.1135
30. Frank K, Stevens Q (eds) (2007) Loose space: possibility and diversity in urban life. Routledge, New York
31. Bishop PD, Chertok F, Jason LA (1997) Measuring sense of community: beyond local boundaries. J Primary Prev 18(2):193–212. https://doi.org/10.1023/A:1024690424655
32. McMillan DW (1996) Sense of community. J Commun Psychol 24(4):315–325
33. Semenza JC, March TL (2008) An urban community-based intervention to advance social interactions. Environ Behav 41(1):22–42. https://doi.org/10.1177/0013916507311136
34. Lara-Hernandez JA (2023) Emergent temporary appropriation versus lockdown regulations: the case of Venice. In: Sustainable cities and society, recent trends on sustainable smart cities: proceedings of IEEE ISC2 2022. https://doi.org/10.1016/j.scs.2023.104854
35. Hall ET (1966) The hidden dimension. Anchor Books and Colophon
36. Wilkerson A, Carlson NE, Yen IH, Michael YL (2012) Neighborhood physical features and relationships with neighbors: does positive physical environment increase neighborliness? Environ Behav 44(5):595–615. https://doi.org/10.1177/0013916511402058
37. Winstanley A, Thorns DC, Perkins HC (2003) Nostalgia, community and new housing developments: a critique of new urbanism incorporating a New Zealand perspective. Urban Policy Res 21(2):175–189. https://doi.org/10.1080/08111140309952

38. United Nations (2015) Transforming our world: the 2030 Agenda for Sustainable Development. In: Division for sustainable development goals, New York, NY, USA
39. Berkes F, Colding J, Folke C (2003) Navigating social-ecological systems: building resilience for complexity and change. Cambridge University Press. http://search.ebscohost.com/login.aspx?direct=true&db=nlebk&AN=120253&site=ehost-live
40. Gershenson C (2007) Design and control of self-organizing systems. Vrije Universiteit Brussel
41. Nyström M, Folke C (2001) Spatial resilience of coral reefs. Ecosystems 4(5):406–417. https://doi.org/10.1007/s10021-001-0019-y
42. Lara-Hernandez JA (2023) Building urban resilience through infrastructure exaptation. In Allam Z, Chabaud D, Gall C, Pratlong F, Moreno C (eds) Resilient and sustainable cities, no 1. Elsevier, pp 537–549. https://doi.org/10.1016/B978-0-323-91718-6.00021-9537
43. Marcus L, Colding J (2014) Toward an integrated theory of spatial morphology and resilient urban systems. Ecol Soc 19(4). https://doi.org/10.5751/ES-06939-190455
44. Carta M (2021) No more masterplan! Resilient communities require incremental, adaptive and generative processes. In: Carta M, Perbellini M, Lara-Hernandez JA (eds) Resilient communities and the Peccioli Charter—towards the possibility of an Italian charter for resilient communities. Springer Nature.
45. Kwok AH, Doyle EEH, Becker J, Johnston D, Paton D (2016) What is 'social resilience'? Perspectives of disaster researchers, emergency management practitioners, and policymakers in New Zealand. Int J Disast Risk Reduct 19:197–211. https://doi.org/10.1016/j.ijdrr.2016.08.013
46. Pickett STA, Cadenasso ML, Grove JM (2004) Resilient cities: meaning, models, and metaphor for integrating the ecological, socio-economic, and planning realms. Landsc Urban Plan 69(4):369–384. https://doi.org/10.1016/j.landurbplan.2003.10.035
47. Gevorkyan AV, Gadhoke P, Freiberg T (2020) A path to global social and economic resilience. August, 0–5
48. Liang J, Li Y (2020) Resilience and sustainable development goals based social-ecological indicators and assessment of coastal urban areas—a case study of Dapeng New District, Shenzhen, China. Watershed Ecol Environ 2:6–15. https://doi.org/10.1016/j.wsee.2020.06.001
49. Cutter SL, Ash KD, Emrich CT (2014) The geographies of community disaster resilience. Glob Environ Chang 29:65–77. https://doi.org/10.1016/j.gloenvcha.2014.08.005
50. Pelling M (2003) The vulnerability of cities: natural disasters and social resilience. Earthscan
51. Lara-Hernandez JA, Melis A, Lehmann S (2018) Temporary appropriation and social sustainability: the case of the historic centre of Mexico City. APRU Sustain Cities Landsc. https://apru.org/
52. Lara-Hernandez JA, Khemri MY, Melis A (2021) Understanding temporary appropriation and the streetscape design: the case of Algiers, Auckland and Mexico City. In: Di Raimo A, Lehmann S, Melis A (eds) Informality through sustainability: urban informality now. Routledge
53. Therrien MC, Usher S, Matyas D (2020) Enabling strategies and impeding factors to urban resilience implementation: a scoping review. J Conting Crisis Manage 28(1):83–102. https://doi.org/10.1111/1468-5973.12283
54. Kennett DJ, Masson M, Lope CP, Serafin S, George RJ, Spencer TC, Hoggarth JA, Culleton BJ, Harper TK, Prufer KM, Milbrath S, Russell BW, González EU, McCool WC, Aquino VV, Paris EH, Curtis JH, Marwan N, Zhang M et al (2022) Drought-induced civil conflict among the ancient maya. Nat Commun 13(1):1–10. https://doi.org/10.1038/s41467-022-31522-x
55. Mahdzar SSBS (2008) Sociability vs accessibility urban street life. University College London
56. Roever S (2020) Street vendors and public space: essential insights on key trends and solutions. In: Women in informal employment: globalizing and organizing. https://www.wiego.org/sites/default/files/resources/file/StreetVendorsandPublicSpace-AnEBook.pdf
57. Whyte WH (1980) The social life of small urban spaces. In: Common ground. Conservation Foundation
58. Allam Z, Dey G, Jones D (2020) Artificial intelligence (AI) provided early detection of the coronavirus (COVID-19) in China and will influence future urban health policy internationally. AI 1(2):156–165. https://doi.org/10.3390/ai1020009

59. Burroughs S (2017) Development of a tool for assessing commercial building resilience. Proc Eng 180:1034–1043. https://doi.org/10.1016/j.proeng.2017.04.263
60. Ortega-García C (2015) Análisis del cambio en el ámbito comercial. In: Departament d'Urbanisme i Ordenació del Territori. Universitat Politècnica de Catalunya (ed) UPCommons. Facultad de Arquitectura. Universidad de la República, p 16. http://hdl.handle.net/2117/81110
61. Francis M (1984) Mapping downtown activity. J Arch Plan Res 1(1):21–35. http://www.jstor.org/stable/43028671
62. Groat L, Wang D (2013) Architectural research methods, 2nd edn. Wiley
63. Boeing G (2021) Street network models and indicators for every urban area in the world. Geogr Anal 54(3):519–535. https://doi.org/10.1111/gean.12281
64. DeLanda M (2016) Assemblage theory. Edinburgh University Press
65. Lima JJ (2001) Socio-spatial segregation and urban form: Belem at the end of the 1990's. Geoforum 32(4):493–507. https://doi.org/10.1016/S0016-7185(01)00019-7
66. Hubbard P, Kitchin R, Bartley B, Fuller D (2002) Thinking geographically. Continuum
67. Brenner N (2005) New state spaces: urban governance and the rescaling of statehood. Oxford University Press
68. Choay F (2004) El reino de lo urbano y la muerte de la ciudad. Lo Urbano En 20 Autores Contemporáneos 33:23–33
69. Boeing G (2018) Measuring the complexity of urban form and design. Urban Des Int 23(4):281–292. https://doi.org/10.1057/s41289-018-0072-1
70. Brown F, Crompton A (2007) The double fractal structure of Venice. In: Proceedings, 6th international space syntax symposium, İstanbul, 2007, January 2007, pp 11.1–11.12
71. Stanislawski D (1947) Early Spanish town planning in the new world. Am Geogr Soc 37(1):94–105
72. The World Bank (2024) Population density (people per sq. km of land area)—Hong Kong SAR, China. Data. https://data.worldbank.org/indicator/EN.POP.DNST?locations=HK&most_recent_value_desc=true
73. Parvin A, Ye AM, Jia B (2007) Multinivel pedestrian movement: does visibility make any difference? In: 6th international space syntax symposium, vol 40, pp 1–16
74. Relph E (1976) Place and placelessness. In: London: Pion, 2nd edn, vol 1. Pion Limited
75. Cheshmehzangi A, Heat T (2012) Urban identities: influences on socio-environmental values and spatial inter-relations. Proc Social Behav Sci 36(June 2011):253–264. https://doi.org/10.1016/j.sbspro.2012.03.028
76. Lara-Hernandez JA, Melis A, Lehmann S (2019) Temporary appropriation of public space as an emergence assemblage for the future urban landscape: the case of Mexico City. Future Cities Environ 5(1):1–22. https://doi.org/10.5334/fce.53
77. Government of New Zealand (2013) Census map—QuickStats about a place. StatsMaps. http://archive.stats.govt.nz/StatsMaps/Home/Maps/2013-census-quickstats-about-a-place-map.aspx
78. Government of New Zealand (2018) Te Ara encyclopedia of New Zealand. Auckland Places. https://teara.govt.nz/en/auckland-places/page-10
79. Siksna A (2006) The study of urban form in Australia. Int Seminar Urban Form 10(April):89–100
80. Gu K (2010) Urban morphological regions and urban landscape management: the case of central Auckland, New Zealand. Urban Des Int 15(3):148–164. https://doi.org/10.1057/udi.2010.4
81. Dürr E (2011) To belong in aotearoa new zealand: Latin American migrant experiences in multicultural Auckland. J Ethn Migr Stud 37(3):503–519. https://doi.org/10.1080/1369183X.2011.521359
82. Pearson S, Kothari S (2007) Menus for a multicultural New Zealand. Contin J Media Cult Stud 21(1):45–58. https://doi.org/10.1080/10304310601103950
83. Nielsen IQ (2023) Perceptions of living in Auckland's city centre: 2022 survey of residents (Issue January). https://www.knowledgeauckland.org.nz/media/2552/perceptions-of-living-in-auckland-s-city-centre-2022-survey-of-residents-nielseniq-jan-2023.pdf

84. Jacobs J (1961) The death and life of great American cities. Random House
85. Lara-Hernandez JA, Khemri YM, Melis A (2020) Understanding temporary appropriation and the informal use of public spaces: the case of Algiers, Auckland and Mexico City. In: Di Raimo A, Melis A, Lehmann S (eds) Informality now: revisiting the issue of informal settlements and their link to sustainability. Taylor & Francis Group, Routledge

Chapter 4
Assessing the Position of the Neighbourhood-Oriented Management Approach in the Economic Management of Medium-Sized Cities

Hadi RezaeiRad and Behdad Beyranvandnezhad

Abstract Since the neighborhood is considered the most fundamental factor of a city and the citizens of each neighborhood have a common sense of national and social identity in that neighborhood, without a doubt, neighborhood-oriented management plays an essential role in urban management. On the other hand, the economic factor is considered as a crutch in the hands of city managers, so the economic management of medium-sized cities will be closely related to the citizens of the neighborhoods. But unfortunately, nowadays, citizen orientation has become a less important issue in the management of cities. The present study was conducted with the aim of measuring and evaluating the attention of managers to the potentials of urban and neighborhood situations in medium-sized cities (such as Bahar) in the economic management of these cities. Also, measuring and evaluating the strengths and weaknesses of such an attitude has been taken into consideration. The method of the current research is descriptive and analytical, and a questionnaire tool was used to collect the data required for analysis. The statistical population includes managers and experts in the field of urban management in Bahar, which was determined by using Morgan's formula of 196 people. The results of the research indicate that urban management in the common form today does not follow the optimal management of cities because the economic issues and problems of cities are very complex. Therefore, paying attention to the neighborhood-oriented management approach for the development and economic growth of cities and optimal management Their economy is essential as a connection point between the local community and the central government. The results of the single-sample t-test indicated that the component of economic transparency (score 3.25) as one of the components of urban management has the most importance because in recent years, useful efforts have been made by Bahar Municipality to inform the public about the municipality's activities. Also, the average of the central neighborhood with the statistical average (3) shows the average and appropriate level of the central neighborhood of this city in Hamadan province.

H. RezaeiRad (✉) · B. Beyranvandnezhad
Department of Urbanism, Faculty of Art and Architecture, Bu-Ali Sina University, Hamedan, Iran
e-mail: H.rezaeirad@Basu.ac.ir

G. Siew et al. (eds.), *Diversity as Catalyst: Economic Growth and Urban Resilience in Global Cityscapes*, Urban Sustainability, https://doi.org/10.1007/978-981-97-8776-0_4

Keywords Neighbourhood-oriented management · Economic management · Medium-sized cities · Municipality's activities · Bahar city

4.1 Introduction

Cities, like living organisms, have life, and if the disruptive factors to their ecological balance are not controlled over time, the grounds for their degradation emerge [12]. Nowadays, due to the extensive dimensions and nature of urban issues, their complexity has doubled [37]. Also, unprecedented urban growth over the past century has led to various effects on urban environments [68]. Consequently, urban planners and policymakers in recent years have endeavored to address this problem and enhance urban management by adopting a neighbourhood-oriented approach and harnessing connecting social capital [29]. This is because the neighbourhood is the fundamental element of the city and the intermediary link between the city and its residents, holding a special position in shaping and organizing urban affairs [24]. Therefore, neighbourhood management represents a relatively new approach to improving public services [60], as citizen participation and grassroots organizations play a key role in enhancing quality of life and urban development. Without citizen involvement and the use of indigenous knowledge in urban decision-making and planning, urban development does not take place in a desirable manner [26].

On the other hand, citizenship has emerged as a social and political process in civil societies since the nineteenth century, representing a serious interaction between the state and the people. Simultaneously with the issue of membership in nation-states, citizenship has become one of the important social and political issues. Citizenship is perceived as the most dynamic social concept in modern society. In other words, it can be said that the foundation of civil society is formed around the system of social institutions of citizenship. In general, citizenship is a kind of mutual social contract and a series of reciprocal rights between "government over people" and "people over government", as well as a common public sentiment towards national and social identity within a defined scope [62]. This approach necessitates restructuring urban management and facilitating citizen participation in various dimensions.

New urban management, in the form of city councils, neighbourhood councils, and with a focus on neighbourhoods, should understand the local needs of citizens and raise awareness of higher levels beyond their demands. In other words, it should both align with citizens and assist city managers in meeting the needs of city residents effectively, enabling citizens to intervene more effectively in their own destiny and providing a conducive environment for their satisfaction [20]. Additionally, issues such as the inconsistency between participations and municipal decisions, and the ambiguity of the influential role of participation in urban affairs, bring forth numerous challenges in the path of local development. Resolving these issues, other than through increasing citizen participation in areas such as urban management and planning, neighbourhoud-oriented preservation, and investment, and diversifying municipal management from public-executive functions, is not feasible. Therefore, this research aims to examine the necessities and functions of neighbourhood-oriented and neighbourhood participation in urban management at the micro level in Bahar city, a subdivision of Hamedan province.

Considering that the most important and vastest asset of any society is its human resources, there are specific methods for utilizing them. Participatory democracy is one of the psychological and mental needs and a source of respect for humanity. Human participation in determining the destiny of their society and decision-making on matters related to their lives has always existed and been favored at various levels. This need has led governments over the years to delegate a significant portion of their duties in decision-making to citizens, which is considered a natural right in collective life and has been seriously emphasized in recent decades in urban management [79]. Therefore, given the importance of identity or, in other words, neighbourhood-oriented as one of the most fundamental issues in various cultural and social domains in modern societies, especially in countries like Iran, it seems crucial [23]. Hence, the main objective of the present research is to assess the attitudes of officials and urban managers in Bahar county towards the potentials and capacities of neighbourhood-oriented in this county. Additionally, examining the strengths and weaknesses of this attitude has been considered. Based on this, the question arises: What factors determine the implementation of neighbourhood-oriented management and the factors influencing deepening citizen participation in urban management in medium-sized cities like Bahar?

4.2 Background and Theoretical Content

4.2.1 Urban Management

Urban management, synonymous with local governance, entails assuming responsibility for improving the performance of a city. This term is used for various issues related to planning, governance, and urban environments [19]. In essence, it is a key element for enhancing sustainable growth and effective progress through innovation [76].

4.2.2 Social Capital

Social capitals in each neighbourhood are resources accessible through membership in a local community [58]. Places with stronger levels of social capital benefit from a better quality of life because:

- Social capital entails better communication, as individuals spend more time in social organizations.
- There are more volunteers.
- More citizens participate in voting.
- Individuals spend more time socializing with family, friends, and neighbors [66].

4.2.3 Social Trust

Social trust is rooted in trust-building among human target groups [17]. Trust is one aspect of all human relationships and fosters participation and cooperation among community members. Trust accelerates participation in various economic, political, social, and cultural realms and increases individuals' willingness to collaborate with different groups in society. Many social philosophers such as Hobbes and Tocqueville believe that trust is the foundation of social order, reducing tensions and enhancing cohesion. It should be noted that trust is acquired and emerges during socialization and is strengthened through social experiences [7].

4.2.4 Local Community

The term "local community" is used both theoretically and practically to refer to a range of social groups. In a pervasive sense, a local community refers to cohesive groups of individuals who share a set of values and a way of life and are recognized as a group. Beyond this, the concept of this term has two distinct applications: first, to describe existing groups and institutions; and second, to discuss social ideals. In urban studies, this term is used to describe a neighbourhood or a local scale [55].

4.2.5 Social Resilience

Resilience is used to demonstrate the process of adapting to conditions when facing disruptive events. This concept relates to achieving balance and stability in dealing with various challenges and quickly rebuilding and recovering after crises and related incidents [49]. Regarding society and from a social perspective, resilience is associated with factors such as the community's ability to create strong social networks, increase levels of trust and social cohesion, gather and utilize local and international resources, establish responsive social policies and programs, and enhance community awareness and education in various social domains, all of which can help strengthen social resilience. Social resilience can be developed in confronting various challenges, including natural, economic, social, and political crises, structural deficiencies, and unforeseen transformations. For example, if a community has strong and dynamic social networks, individuals can easily share information, distribute resources, and collaborate with each other in facing challenges. This ability to cope with challenges and recover after crises is of paramount importance and is recognized as a key factor in sustainable development and social growth [75].

4.2.6 Participation

The term "participation" linguistically refers to involvement and gathering for a specific purpose. In another definition, participation is the mental and emotional engagement of individuals in group situations that motivates them to assist each other in achieving group goals. Participation and engaging in participation are conscious and self-motivated actions of individuals within a community towards the development plans of the society [82]. The most influential factors affecting individuals' participation include awareness, education, and motivation. Awareness alone is not influential because the result of awareness should be a change in attitude, mood, and behavior [3] (Table 4.1).

Table 4.1 The concept of participation from various perspectives

Scholars	Concept	Source
Rummel [70]	Exercise of power	Hosseini and Kameli [31]
Clark and Roberts (1973)	Process	Hosseini and Kameli [31]
Gaetjens (1986)	Social, general, integrated, multidimensional, multicultural process	Azkiya and Ghafari [9]
Alain [4]	Active participation in groups, belonging to a particular group, and having a stake in its existence	Azkiya and Ghafari [9]
Locke et al. [40]	Decision-making	Hosseini and Kameli [31]
Pearson (1989)	Influence and control	Hosseini and Kameli [31]
Shamaie [73]	Citizens' share in power	Ziari et al. [81]
Pakzad [59]	Transforming an indifferent city dweller into a citizen	Khademalhosseiny et al. [36]
Albrechtslund and Lauritsen [5]	Fearless supervisory conditions	Albrechtslund and Lauritsen [5]
Picore et al. (2014)	Environmental responsibility	Piškur et al. [63]
Adair et al. (2015)	Motivation, influence, and social interaction	Adair et al. [1]
Hammell [28]	Autonomy	Hammell [28]
Mosleh and Larsen [53]	Method of engaging individuals	Mosleh and Larsen [53]

Source The Authors

4.2.7 Neighbourhood-Oriented Management

Neighbourhood management is a process that brings together local communities and local organizations to address local problems and needs, aiming to enhance local services. This process is particularly employed in deprived neighbourhoods. Neighbourhood management represents a relatively new approach to improving public services. At its simplest, neighbourhood management is an interaction between local communities and local service providers at the neighbourhood level to address local problems and improve local services [15]. This type of management is a process that recognizes each neighbourhood as a unique entity, enabling the population living, working, or providing services in the neighbourhood to strengthen their strengths and address new challenges [67]. Therefore, neighbourhood-oriented management is an innovative approach in which the city is divided into several neighbourhoods, and the emphasis is placed on utilizing local human resources and opportunities, with needs assessments being conducted based on residents' statements. In this approach, neighbourhood residents are considered the best individuals to contribute to urban management in order to better understand neighbourhood issues and problems and provide assistance [57]. According to Hutchings [32], neighbourhood management is an approach in which local communities and service providers collaborate to enhance the quality of life and provide facilities at the local level [44] (Table 4.2).

4.2.8 Economic Growth

Economic growth leads to development and increased social production and exchange, ultimately resulting in higher income and improved living conditions for society. This process involves increasing the production of goods and services, creating employment opportunities, enhancing living standards, and boosting investments. Economic growth can occur in two forms: increasing production and increasing productivity. Increasing production refers to expanding the volume of goods and services produced by a society, while increasing productivity indicates improvements in resource utilization and advancements in technology and processes [18].

4.2.9 Urban Economic Management

Enhancing the economic competitiveness of cities within regional urban networks is a primary objective of local governments. Urban management seeks to improve the city's economy vis-à-vis other cities by reinforcing its facilitative role, abstaining from economic intervention, identifying relative advantages, and capitalizing on them. While the local economic environment is influenced by various factors such as

Table 4.2 Perspectives of various scholar regarding neighbourhood-oriented management

Scholars	Perspective	Source
Madanipour [41]	Division of urban space and urban society into smaller units and organizing city administration based on these divisions	Khademalhosseiny et al. [36]
Rafieyan and Hudnson (2005)	Encouragement of local authorities to create and develop collaborative programs among private, local, and governmental sectors to promote development at the local level	Ghaderi et al. [21]
Mahony [43]	Management of resources and current neighbourhood issues through voluntary collective action and community participation	Pourahmad et al. [64]
Hutchings [32]	Collaboration between local communities and service providers to improve the quality of life and provide facilities at the local level	Majedi et al. (2018)
Hamdi and Goethert	Attention to capacity-building factors, empowerment, delegation of authority to citizens in neighbourhood management, and the use of new methods and approaches	Hamdi and Goethert [27]
Meine Pieter Dijk	Changing the role of governments from service providers to empowerment; the necessity of facilitating the participation process and establishing social institutions	Van Dijk [16]
Jacobs [34]	The urban environment and neighbourhoods should be shaped based on local social and cultural interactions. She advocated for the development of active and vibrant neighbourhoods to improve the quality of life for residents	Jacobs [33]

Source The Authors

national and regional conditions, political stability, economic security, and cultural and social factors, the role of urban management in providing infrastructure related to economic development is crucial. It should be noted that improving the business environment in cities increases their contribution to national and regional economies, enabling cities to benefit from broader and more sustainable financial resources and incomes. One of the significant challenges faced by municipalities worldwide is generating sufficient revenue to cover the costs of urban services. In a healthy urban economy, all economic actors pay a fee to the city based on the benefits they derive from urban services, referred to as "levies." [80] In recent years, due to insufficient financial resources and the need to provide services, a significant portion of municipal revenue has come from land (density) sales, which is considered one of the unstable sources.

If urban management intends to operate within the framework of sustainable development principles, it must rely on sustainable revenues, including levies for urban renewal and development. Urban management should formulate urban economic strategies and policies that capitalize on the economic capacities of cities and ensure the ability of cities to foster economic development, wealth creation, and employment. The role of urban management in fostering entrepreneurship and employment is crucial. The prosperity of the urban economy depends significantly on the prominent and vital role of municipalities, and their performance can either contribute to or impede the economic growth of cities. Proper planning in the field of urban economics to capitalize on the geographical, historical, and natural capacities of cities to increase per capita income, entrepreneurship, employment, and income generation is essential because urban economic growth can contribute to municipal revenue generation, and these two issues are directly interrelated [30] Special attention to the formulation and implementation of urban strategies and policies that promote and strengthen the participation of both women and men, optimizing the urban tax system, assisting in the creation of suitable urban jobs, especially for marginalized groups, minorities, women, and youth, will ultimately lead to national economic growth. Local governments, by utilizing permanent urban resources, facilitate the achievement of sustainable revenues.

Considering the spatial structure of cities and their regional position, it is possible to flourish the urban economy by allocating resources and employing a planned management approach to leverage baseline assets. Municipal activities in urban management are not limited to physical interventions and providing infrastructure; rather, they involve planning to create opportunities for local developing businesses and fostering an environment for entrepreneurship, education, and skill enhancement in this realm. Limited urban budgets challenge the attainment of sustainable resources, but incentives for businesses in urban communities can attract capital, and planning to support businesses in the path of prosperity and economic growth can increase per capita income and improve the quality of life. Local economic development is one of the most important strategies for poverty alleviation in many developing countries. The national government sets policies and allocates financial resources, research, and other necessary supports for local economic development to municipalities. Municipalities decide on the implementation and management of strategies to ultimately identify problems, prioritize development projects, and create pathways for development between regions, leading to income-generating opportunities for more individuals, especially the impoverished. Urban development strategies can strengthen local ownership and community participation and leverage local resources and skills.

These strategies enhance development opportunities, including various economic initiatives focused on local development. Municipalities must analyze the city's situation, identify growth opportunities, and choose appropriate strategies to achieve their goals. Development plans should include necessary infrastructure actions for business development and focus on areas with comparative advantages compared to other cities. This development is based on empowering socially and economically disadvantaged groups and contributes to achieving social and economic justice. Municipalities should also create a platform for joint investment between large and small partners [38].

4.2.10 Objectives of Neighbourhood-Oriented Management

According to Beilan (2010), there are six main objectives for neighbourhood-oriented management, which include establishing social security, increasing education, enhancing quality of life, improving neighbourhood health infrastructure, addressing housing issues, and promoting employment [44]. Some of the primary services of neighbourhood management include:

- Core services
- Security, nuisance control, and public surveillance
- Environmental protection
- Coordination of neighbourhood-related services
- Maximizing neighbourhood benefits
- Linkage with local businesses
- Development of local initiatives
- Support for local volunteer groups [65].

In general, the key to the success of neighbourhood-oriented management can be summarized as follows:

- A neighbourhood with a clear and distinct definition and identifiable boundaries
- Engagement and involvement of neighbourhood residents
- A dynamic, influential, and empowered neighbourhood manager
- Local community participation in providing strategic direction
- Support and commitment to local authorities
- Quality information availability
- Commitment from service providers and mechanisms for reconciling resident and service provider interests [15].

In a general summary, it can be concluded that neighbourhood-oriented management is dependent on desirable urban management and local community involvement. When illustrated as intersecting circles, the best scenario is when the three circles overlap, indicating alignment between neighbourhood management, urban management, and local community interests. Conversely, the worst scenario is when these circles have no common intersection points (Diagram 4.1).

4.2.10.1 Research Background

An overview of research conducted in the field of participation indicates the presence of numerous studies on social participation, especially public participation in urban management, highlighting its importance. Coulton [14] conducted a study titled "Defining Neighbourhoods for Research and Policy." In pursuit of the research objective, and considering the assumptions for constructing neighbourhood units that are useful and valid for research, policy-making, etc., conventional administrative boundaries are used. Standardization methods used for defining and measuring

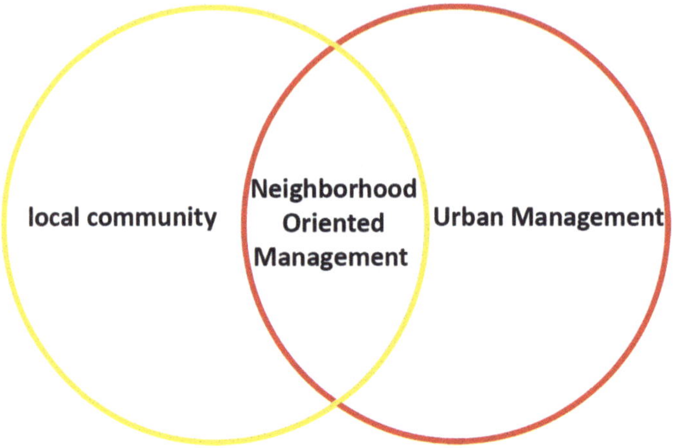

Diagram 4.1 The relationship between influential elements in the emergence of neighbourhood-oriented management

neighbourhoods may become unstable when neighbors perceive neighbourhood space similarly in a nutshell. Geographical Information System (GIS) tools were employed for data analysis, and it was ultimately concluded that attention to small-scale and neighbourhood-based approaches facilitates addressing neighbourhood issues and problems, leading to sustainable neighbourhood development. Lenzi et al. [39] conducted a study titled, "Neighbourhood social connectedness and adolescent civic engagement: an integrative model," examining various modes of youth civic participation in a medium-sized city in Italy.

They developed an integrated model aimed at neighbourhood social development, evaluating the level of youth participation in neighbourhood affairs and intergenerational relationships between youth and adults. They found relatively high levels of youth participation in neighbourhood matters, which contribute to the influence of social systems and the cultivation of youth civic engagement. Maykova and Simonova [47] conducted a study titled "The Participation of Russian Citizens in Local Self-Government: Potential and Real-Life Social Practices," aiming to identify social groups with the potential for social participation in local self-government in Russia. In this research, they highlighted citizen participation as one of the important issues in Russia due to the direct relationship between grassroots and local self-government. They concluded that beliefs in the ineffectiveness of citizens on the activities of local self-government and central government, low trust in the local self-government body, and public unawareness of opportunities for participation in local self-government activities significantly reduce effective citizen participation.

Shi et al. [74] addressed the relationships between various obstacles in the development of sustainable neighbourhoods in China, considering the complex nature of the system. Using an interpretive structural modeling approach, the article proposed a hierarchical model for analyzing the relationships between different obstacles to the development of sustainable neighbourhoods. This hierarchical model, developed using the MICMAC analysis method for classifying factors, examined the complex relationships among various obstacles in the development of sustainable neighbourhoods. Identified barriers with the most significant impact included inappropriate urban planning, lack of supportive policies, lack of national standards and assessment tools, and inadequate planning. They emphasized prioritizing addressing these issues to facilitate the development of sustainable neighbourhoods, aiding in the formulation of urban policies and their implementation.

Bazi et al. [11] in an article titled "Evaluating the role and impact of local councilors on increasing citizens' participation in neighbourhood-based urban management, a case study: Gorgan city" conducted a descriptive-analytical study. They collected data and information through survey methods. The results showed a significant relationship between the performance of local councils and citizen participation in such a way that with a confidence level of 99%, it can be said that increasing and strengthening the performance of local councils will also increase citizen participation in activities related to urban management. Nakano and Washizu [56] investigated the impact of social empowerment on urban structure and smart city development using advanced communication technologies. They employed a hypothesis-based research method for this purpose.

The results indicated that social empowerment through the development of smart cities using advanced communication technologies can have a positive effect on area management and quality of life in cities. Hamdan et al. [26] addressed the impact of collaboration among various organizations in sustainable neighbourhood projects aimed at improving poor performance in energy consumption and greenhouse gas emissions in built environments. The paper examined the challenges and benefits of collaboration among different stakeholders in these projects. Considering the importance of collaboration among diverse stakeholders in sustainable neighbourhood projects, this article identified the influential factors on inter-organizational collaboration through a review of the literature over the past 20 years and proposed solutions to enhance collaboration capacity in these projects.

Mouratidis and Yiannakou [54] investigate factors that make cities livable by using neighbourhood satisfaction and neighbourhood happiness as measures of urban livability. This article employs structural equation modeling methodology. The results indicate that in both cities studied, common factors such as proximity to the city center, neighbourhood safety perception, and spatial coherence are significant determinants of neighbourhood satisfaction. Meanwhile, common factors for neighbourhood happiness include perceived neighbourhood safety, perceived neighbourhood tranquility, neighbourhood social cohesion, spatial coherence, and low neighbourhood density. Moreover, this article highlights significant differences between the two cities under investigation, suggesting that some relationships between neighbourhood characteristics and livability depend on the local context. A summary of research in this field is shown in Table 4.3.

Based on the presented theoretical foundations, the evaluative and conceptual model of the research has been derived as follows (Fig. 4.1).

Table 4.3 Research background and studies related to neighbourhood-oriented management

References	Research title	Findings and results
Coulton [14]	Defining neighbourhoods for research and policy	Emphasising the small-scale and neighbourhood focus facilitates the ease of solving neighbourhood issues and achieving sustainable neighbourhood development
Lenzi et al. [39]	Neighbourhood social connectedness and adolescent civic engagement: an integrative model	Neighbourhood participation systems foster connections and interactions between youth and other generations, nurturing their civic engagement
Maykova and Simonova [47]	The participation of Russian citizens in local self-government: potential and real-life social practices	Belief in the ineffectiveness of citizens on the activities of local self-government and central government, distrust in local self-government bodies, and public unawareness of opportunities for participation in local self-government reduce effective citizen participation
Shi et al. [74]	Challenges of developing sustainable neighbourhoods in China	Complex relationships between various obstacles in sustainable neighbourhood development have been examined, and obstacles with the greatest impact have been identified, such as inappropriate urban planning, lack of supportive policies, insufficient national standards and assessment tools, and inadequate planning
Bazi et al. [11]	Evaluating the role and impact of local councilors on increasing citizens' participation in neighbourhood-based urban management, a case study: Gorgan city	The performance of local councils and citizen participation have a significant relationship, such that with 99% confidence, it can be said that increasing and strengthening the performance of local councils will also increase citizen participation in activities related to urban management
Nakano and Washizu [56]	Will smart cities enhance residents' social capital? The importance of smart neighbourhood management	Social empowerment through smart city development using advanced communication technologies can have a positive impact on area management and quality of life in cities

(continued)

Table 4.3 (continued)

References	Research title	Findings and results
Hamdan et al. [26]	Stakeholder collaboration in sustainable neighbourhood projects—a review and research agenda	The article examines the challenges and benefits of collaboration among various stakeholders in these projects. Considering the importance of collaboration among different stakeholders in sustainable neighbourhood projects, the article identifies effective factors in stakeholder collaboration over the past 20 years of literature review and provides solutions to improve collaboration capacity in these projects
Mouratidis and Yiannakou [54]	What makes cities livable? Determinants of neighbourhood satisfaction and neighbourhood happiness in different contexts	The results show that factors such as proximity to the city centre, neighbourhood security, and spatial cohesion are important in determining neighbourhood satisfaction, while factors contributing to neighbourhood happiness include neighbourhood security, perceived neighbourhood tranquillity, neighbourhood social cohesion, spatial cohesion, and low neighbourhood density

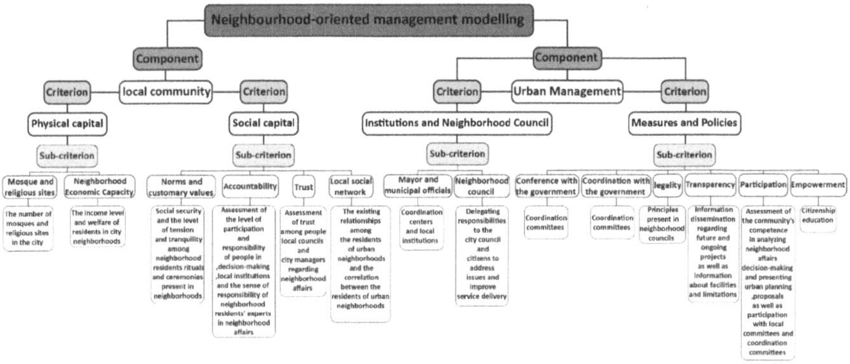

Fig. 4.1 Conceptual model of the research

4.3 Research Methodology

This study employs a descriptive-analytical approach and is applied in nature. Various research methods and data collection techniques, including documentary and field studies, as well as surveys, have been utilized. In order to gather the necessary information for statistical analysis, questionnaires based on the criteria and sub-criteria of urban management and local social components listed in Table 4.4 have been prepared. The validity and reliability of measurements are essential criteria for determining the accuracy and validity of a test. Reliability is associated with random error, while validity is associated with systematic error. Therefore, increasing the sample size can reduce random error and enhance the reliability of the instrument, which in turn affects the accuracy of measurement. Hence, in this study, a statistical population of 400 managers and experts in the field of urban management was considered, and a sample of 196 individuals was estimated using Morgan's formula. Subsequently, quantitative assessments using SPSS software were employed to examine the levels of urban management and local social management as the two main factors in achieving the research objective.

4.4 Study Area

The study area is the Bahar district in the northwest of Hamadan city, initially inhabited by Baharlu residents. Bahar County, with Bahar city as its center, is located in the western part of Hamadan province. It is bordered to the north by Kabudarahang County, to the south by Tuyserkan County, to the east by Hamadan County, to the southwest by Asadabad County, and to the west by Qorveh County in Kurdistan province. This city is situated in the northwest of Hamadan city, approximately eleven kilometers away from it [72] (Fig. 4.2).

4.5 Discussion and Findings

The present research results encompass descriptive and inferential statistics. Descriptive statistics are utilized to succinctly depict the characteristics of the study subject matter in statistical language.

In the current study, the total number of samples was 196, consisting of urban management professionals. This number was calculated using the Morgan formula. Among these, 80% were male and 20% were female.

Based on Table 4.5, it can be observed that the average score for urban management in Bahar County is 3/09 $\pm$ 0/19, while the average score for local community is 2/95 $\pm$ 0/15 The scores of the variables related to urban management and local community indicate the greater importance of urban management components from the perspective of local authorities. This can serve as a determining factor and initiator of urban management in the process of achieving neighbourhood-oriented development. With this interpretation, the components related to the local community can be categorized as physical and social capital, serving as the foundation for realizing this approach, while urban management plays the role of initiator and driving force for initiating and sustaining the utilization of neighbourhood assets.

Based on Table 4.6, which pertains to the inferential statistics of the research, an examination of urban management in Bahar City has been conducted. The results are as follows.

In this study, urban management has been assessed with components and criteria such as empowerment, transparency, legality, participation, etc., reflecting that urban

Table 4.4 Components and criteria of research

Component	Criterion	Sub-criterion		References
Urban management	Policies and measures	Empowerment	Civic education	Naderi Bovanlou and Partovi [55]
		Transparency	Information about future and ongoing projects, Information about facilities and limitations	[55]
		Participation	Evaluation of the public's participation in neighbourhood matters, Local organizations and the level of responsibility felt by neighbourhood experts	Pourahmad et al. [64], Alipour et al. [6], Mohseni and Naemaie [50], Ghafari Gilandeh et al. [22], Mashhadi et al. [46], Naderi Bovanlou and Partovi [55], Valenzuela et al. [78], Baldwind and King [10], Cho and Križnik [13], Lenzi et al. [39] and McGee et al. [48]
		Legalism	Principles in neighbourhood councils	Naderi Bovanlou and Partovi [55]
		Coordination with government	Coordination committees	Naderi Bovanlou and Partovi [55]
		Conference with government	Coordination committees	Naderi Bovanlou and Partovi [55]
	Institutions and neighbourhood Council	Mayor and municipal officials	Coordination centers and local institutions	Naderi Bovanlou and Partovi [55], Alipour et al. [6]; Safaian [71] and Jafaranjad Moghadam [35]
		Neighbourhood council	Delegating responsibilities to the city council and citizens to address issues and improve service delivery	Naderi Bovanlou and Partovi [55], Pawlowski et al. [61] and Power [65]

(continued)

Table 4.4 (continued)

Component	Criterion	Sub-criterion		References
Local community	Social capital	Norms and customary values	Social security and the level of tension and tranquility among neighbourhood residents rituals and ceremonies present in neighbourhoods	Safaian [71]. Majedi and Dashtaki [44], Jafaranjad Moghadam [35], Naderi Bovanlou and Partovi [55] and Power ([65]:3)
		Accountability	Assessment of the level of participation and responsibility of people in decision-making, local institutions, and the sense of responsibility of neighbourhood residents' experts in neighbourhood affairs	Safaian [71], Jafaranjad Moghadam [35] and Naderi Bovanlou and Partovi [55]
		Trust	Assessment of trust among people local councils and city managers regarding neighbourhood affairs	Safaian [71], Adibzadeh et al.[2], Jafaranjad Moghadam [35], Naderi Bovanlou and Partovi [55] and Power [65]
		Local social network	The existing relationships among the residents of urban neighbourhoods and the correlation between the residents of urban neighbourhoods	Safaian [71]; Jafaranjad Moghadam [35] and Naderi Bovanlou and Partovi [55]

(continued)

Table 4.4 (continued)

Component	Criterion	Sub-criterion		References
	Physical capital	Neighbourhood economic capacity	The income level and welfare of residents in city neighbourhoods	Safaian [36], Majedi and Dashtaki [44], Jafaranjad Moghadam [35] and Naderi Bovanlou and Partovi [55]
		Number of mosques and religious sites in the city	The number of mosques and religious sites in the city	Safaian [36], Majedi and Dashtaki [44], Jafaranjad Moghadam [35] and Naderi Bovanlou and Partovi [55]

Fig. 4.2 The position of Bahar city in Bahar county

Table 4.5 Descriptive statistics of variables

Variable standard	Deviation ± mean
Urban management	3.09 ± 0.19
Local community	2.95 ± 0.15

management is in a good state. Comparing the sub-components of urban management, the transparency component and proper communication to the public by the

Table 4.6 Urban management status in Bahar city using one-sample t-test

Urban management components	Mean	Standard deviation	t-value	Degrees of freedom	Significance level
Urban management	3.09	0.19	2.917	39	0.000
Empowerment	3.05	0.06	0.530	39	0.000
Transparency	3.25	0.37	4.120	39	0.000
Legality	2.90	0.44	− 1.433	39	0.000
Participation	2.96	0.26	− 0.968	39	0.000
Coordination with government	3.15	0.73	1.290	39	0.000
Conference with government	3.19	0.48	2.490	39	0.000

managers have the highest score of 3/25. Following that, tangible and visible presence of urban management in the neighbourhood is 3/19, coordination of responsible organizations for urban affairs in the neighbourhood is 3/15, and institutions' involvement is 3/12. The reason for this evaluation is the efforts of the municipality of Bahar City in informing the public in the neighbourhood through public gatherings such as mosques and Friday prayers with the presence of urban management officials, as well as environmental advertisements such as installing billboards and banners, etc. However, these communication and information systems also have weaknesses, and considering that transparency and proper communication are the most important criteria evaluated, it is necessary to strive to improve and develop these systems in the form of online platforms, electronic newsletters, electronic boards, etc. Another component of urban management that has been evaluated is the empowerment component, which includes the sub-criterion of civic education, scoring 3/5 compared to the average score. This indicates the relatively good performance of the municipality in this area. However, considering that empowerment and educating citizens for continuous effectiveness and efficiency must be ongoing, it is suggested to update the civic education system using modern technology in the form of an e-learning network. Moreover, prior to its implementation, necessary information about the educational system should be provided to the public. The average mechanism for full and effective participation of neighbourhoods in urban affairs, compared to the average score of 3, has been 2/96, which is approximately equal to the statistical mean. Therefore, it can be said that the evaluation of the competence of public participation in analyzing neighbourhood affairs, decision-making, presenting urban planning schemes, and participating in local councils and coordination committees in Bahar County is average. This indicates a low level of public participation in urban management, suggesting that managers have relatively weak performance in this area. Given the importance of participation in urban management, the most important indicator for achieving a mechanism for full neighbourhood participation in administering affairs is neighbourhood participatory planning. The comparison of the average scores of the items related to the legality index with the current status of

2/90 indicates the lowest score in the evaluation, suggesting that legality in the city is not in a satisfactory condition. To achieve better urban management, municipalities can enhance legality adherence by providing education on citizen rights and ensuring the commitment of management and officials to equality before the law for citizens (Table 4.7).

In this study, local community was assessed using two criteria: social capital and physical capital. The neighbourhood-oriented is approximately equal to the statistical average of 3, indicating that neighbourhood-oriented in the city of Bahar is at a moderate and appropriate level. Social capital refers to a set of relationships, interactions, and social networks that exist among individuals and social groups, fostering warmth and facilitating social interactions. To measure this concept, indicators such as neighbourhood social network, civic participation, trust, norms, and cultural values were used. Among the criteria of the social capital component, civic participation has obtained the highest score. Comparing with the average of the items, the individual abilities of the residents in the neighbourhood with a score of 3/12 indicate that there is a high potential for achieving neighbourhood-oriented.

City managers can achieve neighbourhood-oriented in Bahar by utilizing the individual abilities of the residents and the existing potentials. The average trust level is approximately equal to the statistical mean (3), indicating that the assessment of trust among the people, neighbourhood councils, and city managers in neighbourhood affairs in Bahar County is moderate. Social trust is considered one of the most important indicators of social capital. Trust between cohesive local communities and local institutions is essential for the formation and development of social capital. This can affect the capacity of neighbourhood management to successfully implement development policies. Therefore, after adopting transparency and empowerment policies and establishing neighbourhood-orinted institutions, neighbourhood

Table 4.7 Status of neighbourhood-centricity in Bahar city

Neighbourhood-oriented components	Mean	Standard deviation	t-value	Degrees of freedom	Significance level
Community neighbourhood	2.95	0.15	− 1.915	39	0.000
Neighbourhood social network	2.95	0.44	− 1.713	39	0.000
Community responsibility	3.11	0.18	2.807	39	0.000
Residents' individual abilities	3.12	0.25	2.037	39	0.000
Norms and ethical values	2.91	0.39	− 2.108	39	0.000
Trust	3.2	0.28	0.522	39	0.000
Neighbourhood economic capacity	2.77	0.29	− 3.741	39	0.000
Religious and public places	2.99	0.38	− 0.198	39	0.000

councils, through urban management, can increase trust among the people, neighbourhood councils, and city managers to create suitable neighbourhood-oriented platforms. Neighbourhood social networks are frameworks where social relationships form, essentially serving as the foundation for social cohesion. The average neighbourhood social network is approximately equal to the statistical mean (3), indicating that in Bahar County, the relationships among people in city neighbourhoods and the cohesion between neighbourhood residents are moderate. Compared to other criteria, the average norms and cultural values are also approximately equal to the statistical mean (3), indicating that social security, tension levels, and the tranquility of neighbourhood residents, as well as the existing customs and ceremonies in neighbourhoods, are at a moderate level in Bahar County.

It seems that municipal managers can enhance neighbourhood communication by expanding the interaction of these institutions with local residents and by extending universal cultural and social centers. This can include assistance in organizing social events (such as mourning ceremonies, celebrations, etc.), reducing unnecessary formalities and permits for neighbourhood activities, and so forth, thereby increasing neighbourhood communication networks to facilitate neighbourhood-oriented initiatives. In this study, physical capital is assessed through two criteria: the economic capacity of neighbourhoods and the presence of mosques and religious places. The average economic capacity of neighbourhoods is lower than the statistical mean (3), indicating that the income and welfare of residents in urban neighbourhoods in Bahar County are relatively low. The average presence of mosques and religious places is approximately equal to the statistical mean (3), suggesting that in Bahar County, religious and public spaces are available at a moderate level. Utilizing neighbourhood assets in the form of public spaces, especially religious and respected locations, paying attention to the presence of local elders and leaders, organizing customary ceremonies and gatherings, and involving community groups and organizations can be useful and effective in initiating and advancing the neighbourhood-oriented development process. Leveraging these assets at the outset of the process, despite the lack of formal management infrastructure, can be highly beneficial and supportive.

4.6 Conclusion and Recommendations

The aim of this study was to assess the perspectives of officials and urban managers in Bahar County regarding the potential and capacities for neighbourhood-oriented management in this region. Additionally, it sought to examine the strengths and weaknesses of this perspective through a survey and a combination of documentary and field studies. After reviewing the relevant theoretical foundations, it became clear that conventional urban management is not sufficient to address the increasingly complex issues and problems facing cities. Therefore, various approaches to urban management have been proposed in recent years to achieve optimal management. One of the most important of these approaches is neighbourhood-oriented management, as it serves as a bridge between local communities and central government. Undoubtedly, the implementation of neighbourhood-oriented urban management approach in intermediate cities like Bahar is confronted with challenges such as the absence of formal structures, authorities, technical capacity, and adequate tools within municipal authorities to fulfill the roles of planning, controlling, and guiding the spatial development of urban neighbourhoods.

Additionally, the execution of projects in a non-expert manner, the hastiness of managerial processes, limitations in the authorities and power of councils and municipalities to enforce comprehensive urban management, and opposition from some governmental organizations to their plans, often developed without supervision and public participation, pose significant obstacles. As mentioned above, a single-sample t-test was employed to examine the urban management in Bahar City and the implementation of neighbourhood-oriented management in this city. The results of this examination indicate that transparency emerges as the most crucial component of urban management, with a score of 3/25. This can be attributed to the efforts of Bahar City municipality in disseminating information to the public during public gatherings. Regarding neighbourhood-oriented management, the average score suggests that it is at a moderate and satisfactory level in Bahar City, as it aligns closely with the statistical average (3).

Consequently, it can be inferred that neighbourhood-oriented management in this city is at an acceptable level. These insights provide valuable considerations for future urban planning and management initiatives in Bahar City and similar intermediate urban settings. The results of the current study are somewhat aligned with previous research. For instance, there is some correspondence between this study and Tahmasebizadeh's research [77], as both studies highlight similar limiting factors in neighbourhood participation, such as the influence of urban management on national decisions. Similarly, there is concordance with Maleki et al. study [45], as both investigations focus on transparency as a key aspect of neighbourhood-oriented management. Additionally, the findings of both Saremi and Mahmudi's research [42] and the present study suggest suitable neighbourhood-oriented management in cities like Aligudarz and Bahar. Furthermore, there is partial alignment with Mashhadi et al. study [46], indicating that residents of Shakib neighbourhood in District 14 of Tehran exhibit a higher-than-average inclination towards participation in urban affairs. It is worth mentioning that there is agreement with Momeni et al.'s research [52, 51] regarding the role of neighbourhood councils as bridges between citizens and urban management.

Given the issues raised in the discussion and the presented solutions, fostering a neighbourhood-oriented mindset among the residents and urban management officials in Bahar city, as well as achieving operational neighbourhood-oriented initiatives, necessitates development, revitalization, and improvement. Therefore, the following proposals are suggested based on the relevant criteria in Table 4.8.

Table 4.8 Solutions and suggestions

Criteria	Sub criterion
Urban management	**Empowerment:** Development, updating, and diversification of communication and information systems **Transparency:** Strengthening communication and information systems of urban management **Legality:** Formulation of regulations for the establishment of neighbourhood councils **Participation:** Strengthening participatory approaches to local institutions **Coordination with the government:** Holding neighbourhood meetings with the presence of urban management officials **Conference with Government:** organizing seminars with urban management officials focusing on the neighbourhood scale
Neighbourhood-oriented management	**Neighbourhood Social Network:** Strengthening the neighbourhood social network through increased social and civic participation **Community Responsibility:** Considering specific responsibilities for residents living in the neighbourhood **Individual Capabilities of Residents:** Identifying potential potentials for use in various areas **Norms and Cultural Values:** Respecting cultural values in the preparation and implementation of plans and programs **Trust:** Using various trust-building methods such as transparency in plans and programs **Neighbourhood Economic Capacity:** Providing a portion of the necessary resources by identifying potential potentials in the neighbourhood **Religious and Cultural Spaces:** Using spaces for gatherings and information dissemination **Neighbourhood Social Network:** Strengthening the neighbourhood social network through increased social and civic participation **Community Responsibility:** Considering specific responsibilities for residents living in the neighborhood **Individual Capabilities of Residents:** Identifying potential potentials for use in various areas

References

1. Adair B, Ullenhag A, Rosenbaum P, Granlund M, Keen D, Imms C (2018) Measures used to quantify participation in childhood disability and their alignment with the family of participation-related constructs: a systematic review. Dev Med Child Neurol 60(11):1101–1116
2. Adibzadeh B, Tafreshi H, Hosseini E (2010) Social engineering and the role of community participation to increase feasibility in renewal of dilapidated urban areas (the case of Nematabad, 19th District of Tehran). J Arch Urban Plan 2(4):141–158 (In Persian)

3. Afshar F, Abbaspour M, Lahijanian A (2020) Evaluation of affecting factors on citizen participation in comprehensive waste management (case study of Districts 8 and 12 in Tehran). Environ Energy Econ Res 4(2):155–164
4. Alain B (1966) Practical social sciences vocabulary. GEditions GEconomy and humanism. [In French]
5. Albrechtslund A, Lauritsen P (2013) Spaces of everyday surveillance: unfolding an analytical concept of participation. Geoforum 49:310–316
6. Alipour S, Hashemi SM, Ameri Sharabi M (2019) Providing a cultural model for developing home businesses in tehran based on qualitative approach. Geogr J Territ 16(61):125–138 (In Persian)
7. Anderson B (2023) Integration: a tale of two communities. Mobilities 18(4):606–619
8. Arnstein SR (1969) A ladder of citizen participation. J Am Inst Planners 35(4):216–224
9. Azkiya M, Ghafari GR (2001) An inquiry of the relationship between social trust and social participation in Kashan rural areas. Nameh-ye Olum-e Ejtemai New (17), 3–31 (in Persian)
10. Baldwin C, King R (2018) Social sustainability, climate resilience and community-based urban development: what about the people? Thames, Routledge, Oxfordshire
11. Bazi KH, Mirkatoli J, Zayedli H (2017) Evaluating the role and impact of local councilors on increasing citizens' participation in neighbourhood-based urban management, a case study: Gorgan city. Geogr Dev Ir J 15(46) (in Persian)
12. Bemanian MR, Rezaei Rad H (2012) Assessing factor analyst and Delphi models by use of GIS in recognizing deterioration in urban fabric (case study of Khaksefid in 4th zone of Tehran). Hoviateshahr 6(11):5–16 (in Persian)
13. Cho IS, Križnik B (2017) Community-based urban development. Springer Singapore, New York, US
14. Coulton C (2012) Defining neighbourhoods for research and policy. Policy Dev Res 14(2)
15. Department for Communities and Local Government (2011) Factsheet 6. Neighbourhood Management
16. Van Dijk M (2006) Managing cities in developing countries: the theory and practice of urban management. Edward Elgar, New York
17. Dinesen PT, Schaeffer M, Sondeskov KM (2020) Ethnic diversity and social trust: a narrative and meta-analytical review. Ann Rev Polit Sci 23:441–465
18. Du X, Zhang H, Han Y (2022) How does new infrastructure investment affect economic growth quality? Empirical evidence from China. Sustainability 14(6):3511
19. Engin Z, Van Dijk J, Lan T, Longley PA, Treleaven P, Batty M, Penn A (2020) Data-driven urban management: mapping the landscape. J Urban Manage 9(2):140–150
20. Falanga R, Ferrao J (2021) The evaluation of citizen participation in policymaking: insights from Portugal. Eval Progr Plan 84(2021):101895
21. Ghaderi S, Fazeli M, Pakseresht S (2016) Neighbourhood development strategies: comparing the performance of neighbourhood management in Iran and Britain. J Commun Dev 8(1(14)):57–88 (in Persian)
22. Ghafari Gilandeh A, Musazadeh C, Ahangari N (2014) Grading district based citizen participation in urban management system (a case study of bukan city). J Urban Econ Manage 2(7):81–99 (in Persian)
23. Ghasemi V, Negini S (2011) The study of the impact of neighbourhood texture on social identity focusing neighbourhood identity in Isfahan city. Urban Reg Stud Res 2(7):113–136 (in Persian)
24. Gholami H, Eliasi E (2013) An analysis of the position of the neighbourhood in Islamic Iranian urban planning. In: National conference on architecture, urban planning and sustainable development with a focus on reading Iranian. Identity in architecture and urban planning. Mashhad (in Persian)
25. Giddens A (1984) The constitution of society, Cambridge, England: polity press.
26. Hamdan HAM, Anderson PH, DeBoer L (2021) Stakeholder collaboration in sustainable neighbourhood projects—a review and research agenda. Sustain Cities Soc 68(2021):102776
27. Hamdi N, Goethert R (1997) Action planning for cities, a guide to community practice. Wiley, New York

28. Hammell KW (2015) Quality of life, participation and occupational rights: a capabilities perspective 62(2):78–85
29. Han Y, Chung RYN (2022) The role of neighborhood social capital on health and health inequality in rural and urban China. Prev Med 156(2022):106959
30. Ho Y, Sarkar S, Alizadeh T (2024) City sizes and economic roles: urban typologies within the Australian urban system. Cities 149(2024):104986
31. Hosseini SB, Kameli M (2017) Measurement of citizen participation in urban management desired (case study district 8 in QOM). Urban Manage Stud 9(29):17–26 (in Persian)
71 Hutchings J (2011) Neighbourhood management research report. Birmingham: birmingham city council report 1–53
33. Jacobs J (2011) The death and life of great American cities: 50th modern library, Anniversary edn
34. Jacobs J (1961) The death and life of great american cities. Random house, Inc, New York City
35. Jafaranjad Moghadam J (2012) Neighbourhood oriented in the urban management system of small cities (case example: Aslandoz city—Ardabil province), Master's thesis, urban design and planning, Islamic Azad University, central Tehran branch (in Persian)
36. Khademalhosseiny A, Tahmasebizad F, Mirenayat N (2015) Analysis of citizens' perceptions of importance of district-centered approach in urban management.(a study on district 9 in Esfahan). Environ Based Territ Plan 8(29):153–176 (in Persian)
37. Kookana RS, Drechsel P, Jamwal P, Vanderzalm J (2020) Urbanisation and emerging economies: issues and potential solutions for water and food security. Sci Total Environ 732(2020):139057
38. Lee CM, Erickson P (2017) How does local economic development in cities affect global GHG emissions? Sustain Cities Soc 35(2017):626–636
39. Lenzi M, Vieno A, Pastor M, Santinello M (2013) Neighbourhood social connectedness and adolescent civic engagement: an integrative model. J Environ Psychol 34:45–54
40. Locke EA, Schweiger DM, Latham GP (1986) Participation in decision making: When should in be used? Organizational dynamics: a quarterly review of organizational behavior for professional managers, 14(3)
41. Madanipour A (2004) Can the management and development of cities be based on neighborhoods?. Applied scientific conference on neighborhood development, perspective of sustainable development of Tehran city. [In Persian]
42. Mahmudi AM, Saremi HR (2015) The role of citizen participant in urban management (case study: Aligudarz City). Am J Eng Res 2320-0847
43. Mahony SO (2007) The governance of open source initiatives: what does it mean to be community managed?. J Manag Gov 11(2):139–150
44. Majedi H, Dashtaki N (2018) Assessment effects of neighbourhood management-based on the trust of citizens in urban management (a comparative study of two Tehran neighbourhoods: Tajrish and Shahadat). Hoviateshahr 11(32):93–102 (in Persian)
45. Maleki S, Zarei J, Zadvali Khajeh S (2019) Assessing the desirable urban governance indexes with the neighbourhood-based management approach (case study: Hamadan city neighbourhoods). Geogr Territ Spatial Arrange 8(29):61–74 (in Persian)
46. Mashhadi Z, Sajjadi J, Shahoseni P (2010) Research of people participation situation in city management with pivotal neighbourhood process case study: Shakib neighbourhood in Tehran of zone14. Geography 4(13):167–189 (in Persian)
47. Maykova EY, Simonova EV (2015) The participation of Russian citizens in local self-government: potential and real-life social practices. Int J Econ Finan 5(Special Issue):142–150
48. McGee TK (2011) engagement in neighbourhood level wildfire mitigation and preparedness: case studies from Canada, the US and Australia. J Environ Manage 92(10):2524–2532
49. Mcgill R (2020) Urban resilience—an urban management perspective. J Urban Manage 9(2020):372–381
50. Mohseni RA, Naemaie B (2017) The role of urban management in promoting social engagement with an emphasis on the neighborhood approach: a case study of district eight in Tehran. J Iran Soc Dev stud (JISDS), 9(3(35)):81–94. [In Persian]

51. Momeni M, Shamskooshki H, Javadian M (2011) Application of neighbourhoods council associations in sustainable urban management based on citizen participation. Proc Eng 21:65–71
52. Momeni M, Azizi H, Taghinia M (2011) Urban management based on citizen participation to enhance quality of life in Tehran
53. Mosleh WS, Larsen H (2021) Exploring the complexity of participation. CoDesign 17(4):454–472
54. Mouratidis K, Yiannakou A (2022) What makes cities livable? Determinants of neighbourhood satisfaction and neighbourhood happiness in different contexts. Land Use Policy 112(2022):105855
55. Naderi Bovanlou M, Partovi P (2011) Presenting community-led urban management pattern for Mashhad. J Arch Urban Plan 3(5):63–80 (in Persian)
56. Nakano S, Wshizu A (2021) Will smart cities enhance the social capital of residents? The importance of smart neighbourhood management. Cities 115(2021):103244
57. Neighbourhood Renewal Unit (2006) Neighbourhood management—at the turning point? Office of the Deputy Prime Minister
58. Nieuwenhuis J (2020) Neighbourhood social capital and adolescents' individual health development. Soc Sci Med 265(2020):113417
59. Pakzad J (2007) Articles about urban design. Armanshahr publications. Tehran. [In Persian]
60. Parker G, Wargent M, Salter K, Yuille A (2023) Neighbourhood planning in England: a decade of institutional learning. Prog Plan 174(2023):100749
61. Pawlowski CS, Winge L, Carroll S, Schmidt TB, Wagner AM, Nørtoft KPJ, Lamm B, Kural R, Schipperijn J, Troelsen J (2017) Move the neighbourhood: study design of a community-based participatory public open space intervention in a Danish deprived neighbourhood to promote active living. BMC Public Health 17
62. Pearce J (2022) Civil society. In: Encyclopedia of violence, peace, and conflict, 3rd edn. Academic Press, Oxford
63. Piškur B, Daniëls R, Jongmans MJ, Ketelaar M, Smeets RJ, Norton M, Beurskens AJ (2014) Participation and social participation: are they distinct concepts? Clin Rehabil 28(3):211–220
64. Pourahmad A, Babaei HR, Rafi A (2017) Explaining the role of neighbourhood management on the social sustainability of Tehran city through urban connecting social capitals. Ir Is City Stud 8(31):75–82 (in Persian)
65. Power A (2004) Neighbourhood management and the future of urban areas. A report from overview & scrutiny: centre for analysis of social exclusion, London, pp 1–37
66. Rae Scheffert D, Horntvedt J, Chazdon S (2020) Social capital and our community, Minneapolis and Saint Paul Minnesota, United States, A publication of University of Monnesota extension center for community vitality
67. Report of the Local Services and Community Safety Overview and Scrutiny Committee (2011) Neighbourhood management. Birmingham City Council
68. Rezaei Rad H, Rafieian M (2016) Evaluating the effects of high rise building on urban heat island by sky view factor: a case study of narmak neighbourhood, Tehran. Naqshejahan 5(4):13–22
69. Roberts N (2004) Public deliberation in an age of direct citizen participation. Am Rev Publ Adm 34:315–353
70. Rummel RJ (1968) The relationship between national attributes and foreign conflict behavior. In: Quantitative International Politics, edited by J David Singer. New York: Free Press
71. Safaian L (2013) Neighbourhood development planning with emphasis on social capital (a case study of Abbasqoli Khan Zanjan neighbourhood). In: National conference on architecture, urban planning and sustainable development with a focus on reading Iranian Islamic identity in architecture and urban planning, Mashhad, Iran (in Persian)
72. Salimi Moayed S (2019) Study of the architecture and culture of the native settlement of Bahar City part I: (typology of houses, outer courtyard and their components). Athar 40(3):196–224 (in Persian)
73. Shamaie A, Adinehvand A, Hajizadeh M (2012) Evaluation of municipalities performance based on good urban governance (case study: yasuj city). J Urban Manag Stud, Fourth Year, No. 11. [In Persian]

74. Shi Q, Yu T, Zuo J, Lai X (2016) Challenges of developing sustainable neighbourhoods in China. J Clean Prod 135(2016):972983
75. Simone C, Iandolo F, Fulco I, Loia F (2021) Rome was not built in a day. Resilience and the eternal city: Insights for urban management. Cities110(2021):103070
76. Tache AV, Popescu OC (2020) Innovative decision making tools used in territorial planning and urban management, Urbanism. Arhitectură. Construcţii 11(2):183–190
77. Tahmasebizadeh F (2022) Analysis of facilitators and limiting participation of citizens in neighborhood planning: case study of isfahan city. J Space Place Stud 1401(24), 95–115. [In Persian]
78. Valenzuela RB, Yeo-Chang Y, Park MS, Chun JN (2020) Local people's participation in mangrove restoration projects and impacts on social capital and livelihood: a case study in the Philippines. Forests 11(5):580
79. VanBussel LGJ, DeHaan N, Remme RP, Lof ME, DeGroot R (2020) Community-based governance: Implications for ecosystem service supply in Berg en Dal, the Netherlands. Ecol Ind 117(2020):106510
80. Wang K, Chung CKL, Xu J, Long Z (2024) Reconceiving China's urban economic transition through symbiotic state-firm dynamics: an integrated perspective from urban governance and global production networks. Cities 150(2024):104974
81. Ziari K, Yadollahnia H, Yadollahnia H (2020) Urban management performance analysis with emphasis on good governance indicators from the citizen's perspective (case study: sari city). Res Urban Plan 11(40):1–16 (in Persian)
82. Ziari K, Beyk Mohamadi H, Rezaeikuchi M (2019) Codification of citizen participation pattern in order to increase the success of city management in Shiraz city. Res Urban Plan 10(38):51–62 (in Persian)

Chapter 5
15-Minute Cities: A Systems Thinking Analysis of a New Urban Development Framework

Using an Agent Based Approach (ABA) to Analyse the 15-Minute City Global Framework and Assess Its Effectiveness, Opportunities, Limitations, and Potential as a New Model for Sustainable Urban Development

Dhruv Parag Parekh

Abstract The 15-minute City, an innovative urban model, emerged prominently during the recent COVID-19 pandemic, capturing the attention of urban planners and policymakers. Rooted in principles of proximity, sustainability, and health, this model envisions a post-pandemic world where all essential services are accessible within a 15-minute travel distance (walking or biking) from every household. These services encompass groceries, medical facilities, childcare, education, employment, and other daily necessities, contributing to the creation of smaller neighbourhood cities within cities. While the concept of a well serviced dense urban lifestyle is not novel and has been integral to various cities globally, the 15-minute City framework offers a conceptual shift to enhance city planning and elevate overall quality of life. Serving as a guiding tool, it holds the potential for widespread applicability across diverse geographical landscapes. This research employs a Systems Thinking lens, using an Agent-Based Approach, to comprehensively analyse the '*Conceptual and Global 15-minute City Framework*' and assess its effectiveness, opportunities, and limitations, as a model for sustainable urban development. Furthermore, it envisions a future system of a '*Comprehensive Proximity-Urbanism Framework*' to support the real-world implementation of 15-minute Cities across varied socio-cultural, economic, and political contexts. The analytical tools used include a DSRP Analysis (Distinctions, Systems, Relationships, Perspectives), a POSIWID Analysis (Purpose of the System is What it Does), and a CAS Analysis (Complex Adaptive System) of the system. Adopting a systems thinking approach allows for a methodical identification of blind spots inherent in this conceptual and global framework. This chapter reflects on the recent scholarly discussions surrounding the 15-minute

D. P. Parekh (✉)
College of Architecture, Art and Planning (AAP), Cornell University, Ithaca, NY, USA
e-mail: dpp45@cornell.edu

G. Siew et al. (eds.), *Diversity as Catalyst: Economic Growth and Urban Resilience in Global Cityscapes*, Urban Sustainability, https://doi.org/10.1007/978-981-97-8776-0_5

City framework to critically analyse its real-world feasibility. In addition, to ground the research and analysis in context the research includes two case study cities. The first case study analyses Dubai's proposed 20-minute City initiative, while the second case study examines Barcelona's 15-minute City analysis of its existing urban morphology. By adopting this holistic approach, the chapter aims to contribute to the ongoing discourse surrounding urban planning paradigms, shedding light on the viability and adaptability of the 15-minute City framework in shaping the cities of the future.

Keywords 15-minute Cities · Healthy cities · Proximity-urbanism · Chrono-urbanism · Sustainable neighbourhoods · Systems thinking · Agent based approach

5.1 Introduction

The concept of the 15-minute City was formally introduced by the French urban planner Carlos Moreno in 2016. It centers around people's ability to access essential urban social functions within a 15-minute walking or cycling radius. The idea gained renewed attention and popularity during the pandemic, as the limitations of car-dependent cities became evident, significantly impacting one's quality of life. This concept evolved into a framework for developing future cities, emphasizing the 15-minute accessibility principle with the goal of reducing emissions, improving social equity, health, and well-being—critical factors in determining overall quality of life [9].

The framework was devised with a golden rule: *"to be able to utilize every square meter of the city in multiple different ways"* and built on four critical pillars of density, proximity, diversity, and digitalization. Essential urban social functions were categorized into *"living, working, commerce, healthcare, education, and entertainment"* [1]. Transport modes were limited to walking and cycling, with the framework acknowledging the arbitrary nature of the 15-minute time limit, suggesting it could be 10 or 20 min based on each city's requirements [1]. This time-based framework has been adopted and proposed as a future development strategy in various global cities, including Dubai [5], Melbourne, Singapore, Beijing, Shanghai, and Portland [6]. Conversely, other cities, such as Barcelona, Munich, and Naples, have examined their existing urban morphology to determine pre-existing compliance with such frameworks [7]. Notably, studies indicate that 80% of Dutch urban areas conform to the characteristics of the 15-minute City [12].

Scholarly and professional discussions on the 15-minute City framework currently revolve around its conceptual and global nature, leading to ambiguity in implementation or analyzing existing cities. Criticisms include its capital-intensive nature, heavy reliance on digitalization, requirements for high density, creation of social enclaves, city fracturing, gentrification, and increased cost of living [4]. Additionally, emerging conspiracy theories hint at the use of such frameworks for exerting authoritarian social control [11].

This research employs a Systems Thinking lens, using an Agent-Based Approach, to comprehensively analyse the '*Conceptual and Global 15-minute City Framework*' and assess its effectiveness, opportunities, and limitations, as a model for sustainable urban development. Furthermore, it envisions a future system of a '*Comprehensive Proximity-Urbanism Framework*' to support the real-world implementation of 15-minute Cities across varied socio-cultural, economic, and political contexts. The Agent-Based Approach is 4-step process that comprises of various analytical tools such as a DSRP Analysis (Distinctions, Systems, Relationships, Perspectives), a POSIWID Analysis (Purpose of the System is What it Does), and a CAS Analysis (Complex Adaptive System) of the system. Finally, the outcomes of the analysis are utilised to create a set of recommendations. Adopting a systems thinking approach allows for a methodical identification of blind spots inherent in this conceptual and global framework.

This chapter reflects on the recent scholarly discussions surrounding the 15-minute City framework to critically analyse its real-world feasibility. In addition, to ground the research and analysis in context the research includes two case study cities. The first case study analyses Dubai's proposed 20-minute City initiative [5], while the second case study examines Barcelona's 15-minute City analysis of its existing urban morphology [6]. By adopting this holistic approach, the research aims to contribute to the ongoing discourse surrounding urban planning paradigms, shedding light on the viability and adaptability of the 15-minute City framework in shaping the cities of the future. In sum, the chapter is structured to include a 'Methodology' section that briefly describes the Agent-Based Approach and its application to this study. It is then followed by a 'Results' section that presents the outcomes of the 3-step analysis and provides the final recommendations. Lastly, the 'Discussion' section provides a concluding summary interpreting the analysis, its outcomes, and proposed recommendations.

5.2 Methodology

5.2.1 Why an Agent Based Approach (ABA)?

We live in a world surrounded by Complex Adaptive Systems (CAS) that operate on the fundamental principle of individual agents adhering to simple rules collectively giving rise to emergent behavior. Contrary to earlier beliefs, these systems are "self-organized" and "leaderless" [10]. This inherent capacity for aggregating behavior enables these systems to adapt and evolve. Examples such as flocks of sheep, ant colonies, the Covid-19 pandemic, and even traffic patterns all represent CAS.

The Agent Based Modeling (ABM) method was developed to understand and simulate social and natural systems that are CAS. However, ABM is not always feasible due to the level of specificity required to model complex systems accurately; its construct validity, as it is challenging to model irrational and subjective agent behavior; and its intense requirements for resources, skills, and time. Thus, the Agent Based Approach (ABA) was developed as an alternative "soft" modeling method to study CASs where ABM was not feasible, which is often the case. While ABA was developed primarily for undertaking policy analysis, it can be widely applied across multiple disciplines and systems that exhibit CAS-like properties [10].

The ABA is a stepwise process focused on "understanding the system," eventually leading to recommendations that enable a CAS system to evolve from its "current undesirable state" to a "future more desirable state" [10]. This research aims to use an Agent Based Approach to critically analyze the '15-minute City' framework as proposed by Carlos Moreno and make recommendations that would allow the effective global implementation of this new sustainable urban development framework.

5.2.2 Steps to an Agent Based Approach (ABA)

The steps to undertake an Agent Based Approach (Fig. 5.1) consist of two main components: first, understanding the system, and second, making recommendations. Various analytical tools are employed to complete these two major components. Within the first component, DSRP mapping and analysis play a crucial role in aligning the researcher's mental model with reality, known as the Systems Thinking (ST) Loop [2]. This alignment enables the POSIWID differentiation analysis, focusing on identifying what the system is doing versus what it should do. These two steps together contribute to gaining a comprehensive understanding of the system.

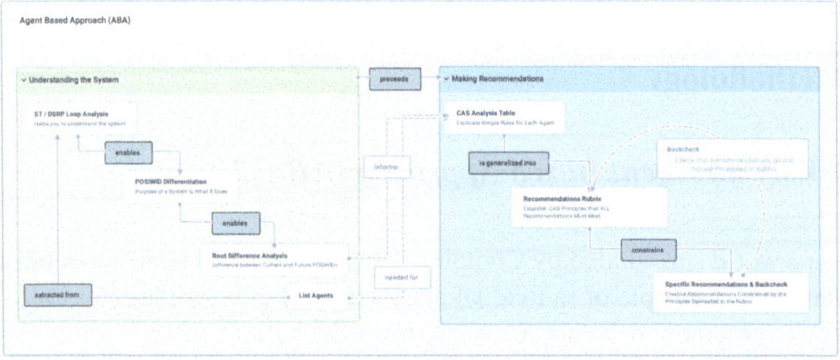

Fig. 5.1 Steps to an agent based approach (*Source* Adapted from the original source, [10])

Next, based on the preceding analysis and identified agents of the system, a CAS analysis is conducted. This analysis classifies the simple rules that govern each agent, revealing potential micro areas of intervention. These rules and agent behaviors are then used to create a recommendation rubric outlining the principles that any recommendations must not violate. Finally, utilizing all the above analyses and criteria, specific recommendations are made and cross-checked against the rubric for compliance [10].

Step 1: DSRP Analysis of the System

The DSRP theory is considered an "atomic structure" in systems thinking. It posits that every system comprises four universal elements, each with two aspects (Table 5.1)—**Distinctions** (identity-other), **Systems** (part-whole), **Relationships** (action-reaction), and **Perspectives** (point-view) [3]. Mapping an existing system using these four universal elements helps paint a realistic picture.

This analysis involves the creation of a DSRP network map based on the DSRP cognitive theory. It is a visual mapping technique that is information-agnostic, triggering an iterative feedback loop to align our mental model with real-world complexities. This mapping method helps one focus on the structure and take an objective view of the system, rather than being engrossed with the complicated nature of available information, thus providing new insights [8].

Step 2: POSIWID Analysis

POSIWID is a Systems Thinking heuristic and acronym developed by scientist Stafford Beer, standing for *"purpose of the system is what it does"* [10]. The key principle of this approach is to accept that the current (often undesirable) results of the system are by design. In other words, "*Instead of the system being badly designed to serve a good outcome, it is brilliantly designed to bring about a bad outcome*" [10]. Following this perspective, a POSIWID Analysis is conducted as the next step within the ABA method. A current-POSIWID and a future should-be-POSIWID are developed and compared to identify structural and root differences within both systems. This process helps gain a deeper understanding of the systems and the factors responsible for their outcomes [10].

Table 5.1 Universal DSRP rules (*Source* adapted from Cabrera et al. [3])

The identity-other distinctions rule	
$D = i \leftrightarrow o$	A distinction (D) is defined as *identity* (i) co-implying an *other* (o)
The part-whole systems rule	
$S = p \leftrightarrow w$	A system (S) is defined as *part* (p) co-implying a *whole* (w)
The action-reaction relationships rule	
$R = a \leftrightarrow r$	A relationship (R) is defined as *action* (a) co-implying a *reaction* (r)
The point-view perspectives rule	
$P = p \leftrightarrow v$	A perspective (P) is defined as *point* (p) co-implying a *view* (v)

Step 3: Complex Adaptive System (CAS) Analysis

The primary objective of a CAS analysis is to identify areas where interventions or recommendations will be most effective. It is based on the principles that CASs consist of multiple micro-agents engaging in simple rules and behavior, leading to macro emergent properties and collective behavior. Thus, understanding these micro-agents and their rules is crucial for successful intervention in an existing CAS. The first step in this analysis is to prepare a list of critical agents for the current and future desired systems, along with defining the problem to be solved. The next step is to identify the simple rules followed by these agents and the corresponding emergent system-level behavior. The results of this analysis will then influence the creation of recommendations in the next step [10].

Step 4: Rubric and Recommendations

The last step involves creating a recommendations rubric followed by the recommendations themselves. The recommendation rubric establishes principles and a framework within which creative recommendations can be made to improve the system. This rubric, along with its principles, is developed from the POSIWID root difference analysis conducted earlier, ensuring the validity and relevance of any proposed recommendations [10].

5.3 Results

5.3.1 DSRP Analysis of the System

The initial step in analyzing the current 15-minute City framework was to conduct a robust DSRP analysis to comprehend the various distinctions, parts and wholes, internal or external relationships to the framework, and the perspectives represented by the framework.

Fropping rules (i.e., framing + stopping rules) [2] facilitated an efficient DSRP mapping exercise, by employing four rules to support this analysis. The primary fropping rule involved identifying and mapping the originally conceptualized 15-minute City framework by Carlos Moreno. Second, since the overall policy framework was conceptual, a critical literature review of its current state, worldwide implementation, and ongoing critiques was undertaken to align it with its real-world presence and included in the map. The expansive literature available then highlighted the need for further analysis from the perspective of a specific city. Thus, the third rule involved conducting two distinct case studies—one proposing a new vision and the other presenting an analytical framework to analyze existing conditions. Finally, the fourth fropping rule acknowledged that the analysis for this global framework would not be all-encompassing, emphasizing the need for structural predictions to map the unknown. All structural predictions have been identified in blue while the key relationships with the 15-minute City framework are mapped in purple. All maps can also be viewed on Plectica at this link: https://www.plectica.com/maps/546GLFCV7 (best viewed in google chrome).

As the map in Fig. 5.2 demonstrates, the 15-minute City framework is an *"RDS—Relationship Distinction System"* [2] representing people's ability to access essential urban social functions. The entire map—'people accessing essential urban social functions within 15 min of walking or cycling'—structurally represents the RDS-Barbell system (Fig. 5.2). The same structure was used to map the two case study cities, understanding the '20-minute City' proposal in Dubai, UAE (Fig. 5.3), and the '15-minute City' analytical framework in Barcelona, Spain (Fig. 5.4). Additional part-whole systems were then identified and included as relationships to this basic RDS-Barbell system, covering historic trends, ongoing critiques, benefits, and commuting standards.

The DSRP analysis of the framework and two case study cities was particularly useful in understanding the distinctions between what was considered (included) versus not considered (not included) within this new popular conceptual and global framework, highlighting major gaps within the system. The mapping exercise also aided in understanding the entire 15-minute City framework, including its part-wholes, relationship loops, and perspectives, along with external factors influencing or catalyzing the creation of such a system.

Structural predictions throughout the process (i.e., fropping rule 1 to rule 4) played a vital role in discovering new questions or linkages ('?') and helped highlight the unknown. A major structural prediction for the global 15-minute City framework was

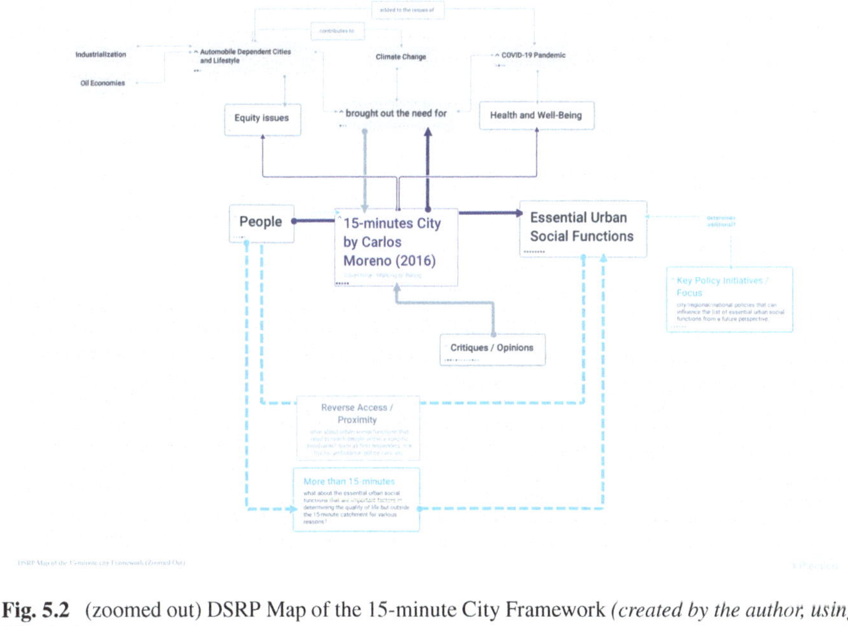

Fig. 5.2 (zoomed out) DSRP Map of the 15-minute City Framework *(created by the author, using Plectica)*

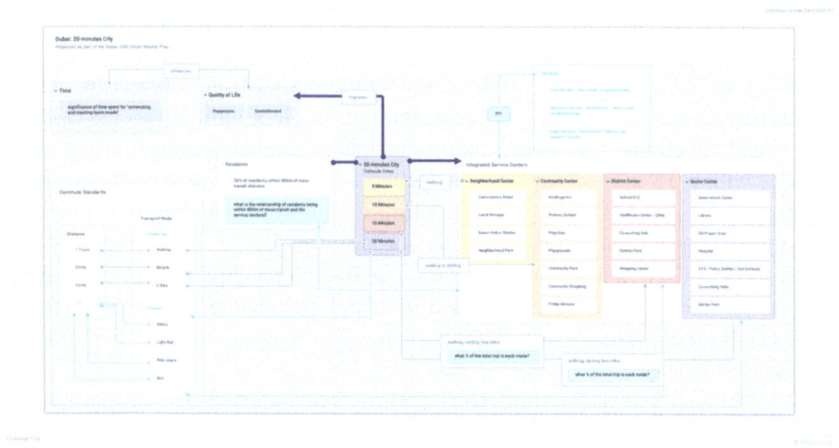

Fig. 5.3 DSRP Map of Dubai's proposed 20-minute City strategy *(created using Plectica)*

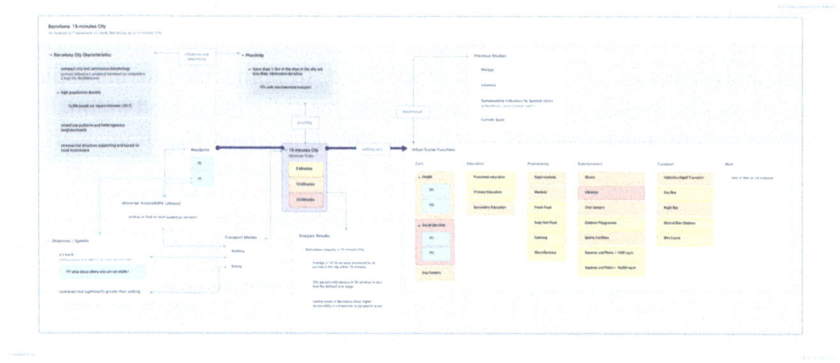

Fig. 5.4 DSRP Map of an analytical framework assessing Barcelona as an existing 15-minute City *(created using Plectica)*

to identify essential urban social functions critical to a person's quality of life but not situated within the 15-minute catchment. Conversely, another significant structural prediction was made to understand the essential urban social functions that need to access a person/house within 15-minutes, such as emergency and safety responders like fire trucks, ambulances, first responders, and police, which are vital services impacting one's quality of life.

Furthermore, the two case study cities helped reinforce the mental model of this framework with real-world implementation, engaging the Systems Thinking (ST) loop. For instance, the Dubai case study highlighted the need to detail out the scale of a facility present within the 15-minute catchment to better understand service provision, as demonstrated by the classification of healthcare into polyclinics, healthcare centers, hospitals, etc. Meanwhile, the Barcelona case study underscored the concept of *"universal accessibility"* wherein walking is considered relatively more accessible than cycling [6].

5.3.2 POSIWID Analysis

The Step 1 DSRP analysis of the existing 15-minute City model, along with the DSRP analysis of two case studies in Dubai, UAE, and Barcelona, Spain, has helped determine the potential future state of this existing system. For Step 2 POSIWID analysis, the current system has been classified as a *"Conceptual and Global 15-minute City Framework"* while the future system has been classified as a *"Comprehensive Proximity-Urbanism Framework"*.

The key root-difference findings between the current and future systems are associated with the fact that the current system is presented as a high-level conceptual idea to influence urban living. On the other hand, the future system is focused on evolving this concept into a strategy that can be implemented in real-world scenarios across distinct and varying geographies. The results of the POSIWID analysis are documented in the Table 5.2.

5.3.3 CAS Analysis

Building on the results of the Step 1 and Step 2 analyses, various agents critical to the creation, implementation, and use of a 15-minute City Framework (current system) or a Proximity-Urbanism Framework (future system) have been identified. While the system may include other agents, the ones listed below are most crucial for grasping the mechanics and rationale behind the system's implementation. These identified agents of the given Complex Adaptive System (CAS) have been grouped into four broad categories: the users, the leaders, the urbanists, and the developers. It is to be noted that this classification is based on a globalized perspective (i.e., using two international case studies) and will need to be modified for a region-specific analysis. The results of the CAS analysis are documented in Table 5.3.

5.3.4 Rubric and Recommendations

Following the Step 3 POSIWID Analysis, a recommendation rubric has been developed. This rubric comprises a set of principles guiding the formulation of policy interventions and recommendations, serving as a mechanism to ensure alignment with the preceding analysis.

Table 5.2 POSIWID root-difference analysis of the conceptual 15-minute City and a proximity-urbanism framework

POSIWID comparison			
Current system	A conceptual and global **15-minute City** framework		
Future system	A comprehensive **proximity-urbanism** framework		
POSIWID analysis			
	Current POSIWID	**Future POSIWID**	**Root Difference**
	The 15-minute City Framework:	*The Proximity-Urbanism Framework:*	*At its core, the POSIWID Root difference is:*
1	Is a global and conceptual model introducing the idea of accessing select urban social functions within a given timeframe (short) to improve quality of life	Is a contextual and comprehensive development strategy prioritizing and structuring proximity to all critical urban social functions to improve quality of life	• Conceptual versus comprehensive • Global versus contextual • Selective urban social functions versus all critical urban social functions
2	Is focused on proposing new principles and an ideal model of urban development	Is focused on implementation of the idea in the grounded socio-cultural, political, economic, and systemic reality of cities	• Ideal versus realistic • Principle versus implementation
3	Utilizes time as the only and fundamental metric of assessment	Utilizes the all-encompassing proximity (time, distance, accessibility for diverse user groups, transport, climate) as a fundamental metric of assessment	• Time driven versus proximity driven
4	Considers the history of urban development, and the present-day challenges but does not account for future evolution/changes	Considers the future way of life enabled through technological evolution (such as but not limited to future modes of transport, future working patterns, future service deliveries and future systems)	• Past and present versus future ready
5	Proposes access via transportation modes that are limited to walking and biking only	Proposes access via an integrated transit strategy that accounts for pedestrians (walking, biking, motorized scooters etc.) and public transit (bus, rail, metro etc.) including future modes of transport	• Pedestrians versus integrated transit
6	Is applicable to high-density urban areas only	Extends its applicability from low-density (sub-urban) to high-density (urban) areas through distinct and separate solutions	Urban density versus urban/sub-urban density

(continued)

Table 5.2 (continued)

7	Proposes maximum time based standards	Proposes relevant best practice guidelines along with minimum/maximum standards to implement proximity	• Maximum limits (standards) versus best practice ranges (guidelines)
8	Catalyzes a capitalistic (profit-making) response in the real estate industry causing gentrification, displacement, and increased costs of living exacerbating equity issues	Creates control mechanisms and interventions to prevent capitalistic responses to the addition of critical urban social functions in existing areas promoting equity and overall quality of life for all	• Ignores downstream equity issues versus Tackles equity issues upfront

Recommendation rubric

No specific recommendation should violate these rules

1. Proposed framework/strategies/criterion must be contextual responding to the ground realities of the said urban area
2. Proposed framework/strategies/criterion must be comprehensive and nuanced to enable successful implementation
3. Proposed framework/strategies must look at access to urban social functions using an all-encompassing metric of proximity for assessment, instead of time alone
4. Proposed framework/strategies must be future ready by taking into consideration technological advancements, changes in lifestyle patterns etc.
5. Proposed framework/strategies must have explored broader alternatives before narrowing down to specifics (example the use of specific modes of transport)
6. Proposed framework/strategies must take into consideration equity and diversity factors impacting the said urban area

Specific interventions/recommendations

Intervention/Recommendation #1: Develop the current Global 15-minute City conceptual framework into a contextual and regional Proximity-Urbanism Framework that focuses on *"proximity"* as a factor of time, distance, mode of transport, climatic conditions and user accessibility to the desired urban social functions

Intervention/Recommendation #2: Assess the ground realities of user patterns and consumer behaviour through robust large data collection, analytics, and civic engagement. The data collection may use innovative technological modes such as but not limited to volunteer movement tracking devices. Furthermore, identify the gaps in current provisions for which data could not be collected and benchmark against target standards and best practices

Intervention/Recommendation #3: Prepare a list of 'all critical' urban social functions necessary to improve quality of life in the said urban area for a diverse population. Based on this list and using integrated modes of transportation, create the hierarchy and structure of provision for each urban social function to acknowledge urban social functions that will be provided within and outside a 15-min (or relevant) catchment

Intervention/Recommendation #4: Develop strategies/solutions that cater to the real need and use. Implementation of the framework and its strategies is key. Successful implementation reflects utilization by the local residents. Recognise the power of agency and everyday choices made by the local population

Table 5.3 CAS Analysis identifying the salient agents involved

List of salient agents	Current simple rules	Current system level behaviour
The users: people/locals/ residents	• Access urban social functions within 15-minute • Use non-vehicular modes of transport (pedestrian/public transit) • Generate demand for services/resources • Consume services/resources • Hold governments accountable	• Successful utilization of a 15-minute City • Improves quality of life • Improves overall well-being and happiness • Reduces carbon emissions • Influences consumer and lifestyle behaviour/habits/ choices Improves economic productivity
The leaders: government agencies/politicians/leadership	• Create a customized vision for a 15-minute City • Develop a contextual 15-minute City model *(with the help of Urbanists)* • Facilitate and oversee implementation of the 15-minute City model (through standards, guidelines, policies, laws, etc.) • Determine funding sources • Understand Public need • Gather public support	• Manage successful implementation of a 15-minute City • Achieve Sustainable Development Goals • Fosters an overall sense of belonging, inclusion and cohesion
The urbanists: spatial planners, designers, architects, policymakers, researchers	• Provide technical support to develop a robust, realistic, and implementable model • Provide support to gather public data and understand choices/behaviours • Provide consultation/ critiques to governments on proposed strategies	• Support in successful implementation of a 15-minute City • Plan, design and develop implementation strategies
The developers: real estate developers/businesses/ corporations	• Promote execution of the proposed 15-minute City model • Create economic incentives for businesses/service providers *(this may sometimes be performed by the governments)* • Build/retrofit the physical environment to enable the 15-minute City	• Execute the 15-minute City

5.4 Discussion

5.4.1 DSRP Analysis of the System

The 15-minute City framework, as examined through DSRP mapping, is a straightforward and conceptual idea aimed at enhancing the quality of life, building urban resilience, reducing emissions, fostering sustainable development, improving health and well-being, and advancing social equity through enhanced accessibility. However, significant gaps and potential blind spots have been identified within the DSRP map, which may undermine the framework's goals.

Firstly, the framework fails to acknowledge or address the diverse groups and roles of people with distinct needs to access urban social functions. This oversight could lead to the formulation of policies and the development of cities that lack social equity. Secondly, concerns arise regarding the implementation of such a framework in existing cities worldwide due to its lack of prescriptiveness. Additionally, the high-level and vague nature of the essential urban social functions, without guidance on policy enforcement, poses challenges. These functions do not encompass every aspect influencing quality of life, and the framework fails to highlight functions critical to quality of life situated outside the 15-min radius due to scale or agglomeration related factors. The framework also overlooks existing or future policy initiatives that should influence the categorization of these functions.

Lastly, the framework limits modes of transport to walking and cycling alone, neglecting universal considerations, and overlooking future personal mobility options (beyond cars), such as e-scooters or public transit. This limitation narrows the framework's approach, preventing it from being holistic in consolidating various urban social functions essential for achieving its developed goals.

5.4.2 POSIWID Analysis

The current POSIWID Analysis clearly highlights the strengths of the existing system in presenting a conceptual idea and propagating it globally. It has garnered attention for this idea and generated momentum to influence new sustainable urban development practices. However, the current system's capacity and role are limited to this global promotion. As discussed within the DSRP mapping, it lacks the necessary structure, detail, and inclusivity required for actual framework implementation, underscoring the need for a comprehensive Proximity-Urbanism framework.

A critical difference identified between the current and future POSIWID lies in the factors determining accessibility and proximity to urban social functions. It's crucial to recognize that accessibility and proximity cannot be defined solely by time but are influenced by time (5 min, 10 min, 20 min, 30 min, etc.), distance (200 m, 500 m, 1000 m, 2000 m, etc.), mode of transport (walking, cycling, e-scooters, public transit, future modes, etc.), climatic conditions (extreme summer, extreme winter, temperate climate zones etc.), and user groups (age, gender, specially-abled, etc.). The analysis also emphasizes that the future framework needs to proactively address social equity issues, be more inclusive, and be contextual to enable successful implementation in existing cities. The root differences identified in the current and future POSIWID analyses present an opportunity to create strong recommendation principles guiding the formulation of recommendations for a future framework.

5.4.3 CAS Analysis

The CAS analysis faces a similar challenge as that of the conceptual 15-minute City framework—it lacks context and has a global nature. The agents identified within the CAS analysis have been categorized into four broad global categories, acknowledging variations across cities, which may not universally apply. This underscores the necessity for a contextual analysis and implementation framework based on the ground realities of any specific city.

Furthermore, the analysis highlights the critical role of users and leaders in ensuring the successful implementation of this framework. It emphasizes the need for employing more bottom-up approaches within this top-down policy framework to enhance its robustness and inclusivity. A bottom-up approach would genuinely allow for considering the diverse needs, perspectives, and aspects of the residents impacted by such a framework. This analysis vividly demonstrates that achieving various quality-of-life and sustainability goals is an emergent property of the successful utilization of such a city by its people. Thus, the effectiveness of this policy framework would be measured by its successful utilization.

5.4.4 Rubric and Recommendations

The analyses conducted within this ABA method have revealed that the successful implementation of any proximity-based framework relies on its nuanced consideration of various groups, factors, and functions. It should be contextual, addressing the on-ground realities of a specific city. Violating any of the rubric principles would lead to a repetition of the conceptual 15-minute City framework, hindering the desired outcomes. The key recommendations derived from this ABA analysis suggest using the current conceptual and global 15-minute City framework as a starting point to propose a comprehensive proximity-urbanism framework specific to each region or city intending to implement this idea. It is imperative that this new framework takes into consideration the everyday practices native to the place and accounts for varying needs and behavior patterns observed through robust data collection processes.

Furthermore, it is critical to define *"proximity"* in each context, not dependent solely on time, and to consider multiple factors influencing residents' proximity and accessibility to essential urban social functions. Lastly, understanding residents' requirements for these functions is essential, rooted in place and varying based on socio-cultural practices worldwide. All the aforementioned recommendations aim to enable a successful implementation of the comprehensive framework by intervening at the agent group of *"the user's"*, as identified in the CAS analysis. Understanding resident practices, behavior, and requirements will help create a strategy that, once implemented, will be efficiently utilized by the residents.

Additionally, while this analysis has not explicitly recommended interventions at the agent group of *"the leaders"*, the CAS analysis suggests potential for proposing strategies that guide or hold leaders accountable in the process.

Acknowledgements I am deeply grateful to my professors, Dr Derek Cabrera and Dr Laura Cabrera, for their exceptional instruction in the Systems Thinking course at the J. E. Brooks School of Public Policy, Cornell University. Their comprehensive knowledge, engaging teaching methods, and unwavering support were instrumental in shaping my understanding of this innovative approach and inspiring my passion for its application in urban planning and design. This study would not have been possible without their guidance and encouragement.

References

1. Allam Z, Moreno C, Chabaud D, Pratlong F (2023) Proximity-based planning and the "15-Minute City": a sustainable model for the city of the future. The Palgrave handbook of global sustainability. Springer International Publishing, Cham, pp 1523–1542
2. Cabrera D, Cabrera L, Cabrera E (2020) An evolving evidence-base and literature review of the universal patterns and atomic elements of complex cognition. Cabrera Research Lab. Ithaca, NY. help.cabreraresearch.org/. Accessed 14 June 2024
3. Cabrera D, Cabrera L, Midgley G (2021) The four waves of systems thinking. In: Cabrera D, Cabrera L, Midgley G (eds) Routledge handbook of systems thinking. Routledge. London, UK.
4. Casarin G, MacLeavy J, Manley D (2023) Rethinking urban utopianism: The fallacy of social mix in the 15-minute city. Urban Stud 00420980231169174
5. Dubai 2040 (2023) The 20-minute City. Dubai 2040 urban master plan. http://dubai2040.ae/en/projects-and-initiatives/the-20-minutes-city/. Accessed 6 Jan 2024
6. Ferrer-Ortiz C, Marquet O, Mojica L, Vich G (2022) Barcelona under the 15-minute city lens: mapping the accessibility and proximity potential based on pedestrian travel times. Smart Cities 5(1):146–161
7. Gaglione F, Gargiulo C, Zucaro F, Cottrill C (2022) Urban accessibility in a 15-minute city: a measure in the city of Naples, Italy. Transp Res Proc 60:378–385
8. Lucio-Maymon P, Castellanos-Canales A, Cabrera L, Cabrera D (2021) The universal cognitive grammar of systems mapping: a rubric to evaluate the various tools and techniques of systems mapping. In: Cabrera D, Cabrera L, Midgley G (eds) Routledge handbook of systems thinking. Routledge, London
9. Moreno C, Allam Z, Chabaud D, Gall C, Pratlong F (2021) Introducing the "15-Minute City": sustainability, resilience and place identity in future post-pandemic cities. Smart Cities 4(1):93–111
10. Mulyono YO, Sukhbaatar U, Cabrera D (2020) "Hard" and "Soft" methods in complex adaptive systems (CAS): agent based modeling (ABM) and the agent based approach (ABA). J Appl Syst Think 20:9

11. Paddison L (2023) How "15-minute cities" turned into an international conspiracy theory | CNN. CNN. https://www.cnn.com/2023/02/26/world/15-minute-cities-conspiracy-theory-climate-intl/index.html
12. Papas T, Basbas S, Campisi T (2023) Urban mobility evolution and the 15-minute city model: from holistic to bottom-up approach. Transp Res Proc 69:544–551

Chapter 6
Urban Horizons in China: Challenges and Opportunities for Community Intervention in a Country Marked by the Heihe-Tengchong Line

Javier Morán Uriel, Federico Camerin, and Rafael Córdoba Hernández

Abstract The research examines the evolution of urbanisation and population density in China between 2000 and 2020, with three key objectives. Firstly, it analyses how state policies have influenced the configuration of the real estate market and urban sprawl in the country. Secondly, it studies the variation of population density through annual census data, focusing on how policies have affected demographic distribution. Thirdly, it investigates the impact of national policies on population density in relation to community policies. The inquiry employs data from over 5000 urban entities to assess the influence on the spatial distribution of population. Historically, the relationship with Chinese community policies is manifested in the promotion of a balanced regional development attempt, the reform of the hukou system to facilitate mobility and rural–urban integration, and, more recently, the implementation of strategies for sustainable urbanisation. These policies are designed to enhance the living standards and urban infrastructure in Chinese cities by addressing the challenges of urban concentration while promoting more inclusive and balanced development across the country. However, the outcomes of these policies are yet to be determined. The chapter comprises the following sections. Section 6.1 outlines the methodology employed in the research, which culminates in an examination of urban development policies in China and their impact at the national level (Sect. 6.2). The research then proceeds to examine the impact of state interventions on urbanisation patterns (Sect. 6.3), before summarising the key findings on the challenges China faces in density distribution for the creation of effective community policies (Sect. 6.4). The implications of these issues are discussed in Sect. 6.5, which leads to a final section that highlights the main lessons learned and provides recommendations for future policies (Sect. 6.6).

J. Morán Uriel · R. Córdoba Hernández
Departamento de Urbanística y Ordenación Territorial, Universidad Politécnica de Madrid, Madrid, Spain

F. Camerin (✉)
Departamento de Urbanismo y Representación de la Arquitectura, Universidad de Valladolid, Valladolid, Spain
e-mail: federico.camerin@uva.es

G. Siew et al. (eds.), *Diversity as Catalyst: Economic Growth and Urban Resilience in Global Cityscapes*, Urban Sustainability, https://doi.org/10.1007/978-981-97-8776-0_6

Keywords Urban development · Real estate market · Urbanisation · Growing economy · Urban sprawl · Density

6.1 Introduction

The research involved three principal tasks. The first of these examines the principal documentary sources and reviews the scientific literature to interpret the urban development policies implemented in recent years in China and the functioning of the Chinese real estate market. The objective is to gain insight into the complex process that has led to the current situation.

The second task ascertains whether the pattern of population density has undergone a transformation over time because of governmental policy. The data employed in this study are primarily derived from the population figures of the National Bureau of Statistics of China, which are derived from the national census conducted every ten years. Additionally, the main data bulletins of the provincial-level administrative units (autonomous regions, municipalities directly under the central government) and the statistical bureaus in China have been utilised. It is also important to note that the two Special Administrative Regions of Hong Kong and Macao, as well as Taiwan, were included in this study. The data on population for Hong Kong and Macau were obtained from the Census and Statistics Department, SAR Hong Kong, and the Direcção dos Serviços de Estatística e Censos, Macau, for the years 2001, 2011 and 2021, respectively. Finally, the data for Taiwan were sourced from the National Statistics, Republic of China.

The resident population database, which is linked to the different units of representation (provinces, prefectures and counties), was constructed for this research from the 2852 spatial units identified for the year 2020. This was achieved by entering the information from the China National Bureau of Statistics and geolocating it in each of the polygonal units that constituted the cartographic base. Each of these spatial units was adequately georeferenced in the aforementioned database but lacked precise population data from the last three censuses and the resulting density. For this reason, the data had to be processed through geospatial processing tools using Geographic Information Systems (GIS) processing tools, which proved to be an ideal system for the processing of information in this phase of the research, given that it allows the collection, management and analysis of geolocalised data. The process yielded a database that could be utilised to illustrate the evolution of density at the three primary administrative levels. The density considered in this database is defined as the number of permanent residents in each spatial unit at a specific time. The selected time points were the years 2000, 2010 and 2020.

The third task ascertains whether Chinese national policies at the city level have had an impact on the country's development model and whether the relationship between density and the imaginary Aihui-Tengchong Line remains consistent. This is an identifying element of the spatial distribution model of the population which separated the Eastern and Western sectors of the country, respectively with a high and low population density. This task is performed on the mapping of the 5047 main urban entities with more than 25,000 inhabitants identified in the "City population dataset" [3]. A database of geolocated entities was created for this purpose based on the entity's name, its relationship to the three main administrative levels existing in China ("1. Province", "2. Prefecture", and "3. County"), and the population of each entity according to the most recent census data available from the China National Bureau of Statistics [29], the Bureaus of Statistics in China [4], the Census and Statistics Department, SAR Hong Kong [6], and the Macao's Direcção dos Serviços de Estatística e Censos [6]. The verification process of the entities contained in this database was conducted by superimposing them on the OpenStreet Maps cartography, which was loaded directly into the ArcGIS Pro program. In addition, the data for each of the cities included in the City population dataset was transcribed into an Excel spreadsheet format to facilitate the integration of the alphanumeric population data with the respective graphic entities.

To study the density of territories, a heat map is created in which each entity represents a surface area, and its relative density dynamics are analysed. This type of cartographic representation is particularly well-suited to the display of point distributions, particularly when the points in question are near one another. The symbology employed in this representation indicates the relative density of points as a dynamic raster visualization, with a colour scheme employed to indicate density values. This task was performed through the geographic information systems tools to the totality of entities suitably geolocated, represented by a colour representation that varies from cold (sparsely populated) to warm (densely populated). As the user zooms in and out, the density definition and, therefore, the colour values change. The density is calculated using the Kernel Density Method, which determines a magnitude per unit area from point or polyline entities [18] in order to fit a smoothly tapered surface to each point.

6.2 China-Led National Policies and Their Impact on the Country's Development Model

The hukou is the census registration system for individuals or households in China and was first introduced in 1958, with a design like the propiska system employed by the Soviet Union at that time. Its continued use in China is essential for understanding the migratory movements of individuals and households from rural to urban areas within the country. The hukou system controls internal migration and limits access to public services for rural migrants and it has contributed significantly to the formation of cities in China and the formalisation of so-called 'ghost cities' [14, 56]. This system has prevented many new cities from reaching adequate population density by restricting the ability of rural citizens to settle in urban areas, resulting in vast uninhabited urban areas.

The hukou divides all Chinese citizens into two distinct categories (yin-yang), differentiating between those residing in urban and rural areas. As observed by Qingwu [32], the hukou system has served to consolidate the country's prevailing socialist system and public interests. However, in the literature, the predominant explanation is that it was initially used to limit and control the movement of the population from the rural to the urban environment, but was later remodelled in 1978 to serve the industrialisation of the country and influencing the selection of emigrants through the restriction of social services related to this system and the distortion of the return to qualification through wage discrimination [8, 64].

In the early 1980s, China initiated a series of reforms that relaxed restrictions on internal migration. This represented a significant shift in policy, allowing Chinese citizens to move more freely within the country for the first time in decades in search of better job or educational opportunities [2]. While this liberalisation did not encompass alterations to the hukou system, impeding individuals from relocating with greater ease, Chinese people were still able to officially alter their place of residence without approval, which still constrained their access to services such as education and medical care in their new places of residence [36, 41, 55] (see Fig. 6.1).

Over time, cities began to develop local policies that allowed migrants to settle on a more permanent or semi-permanent basis. The specific policies enacted by different cities varied, but generally included the relaxation of urban hukou requirements and access to certain public services. These policies significantly improved the quality of life for migrants and encouraged a greater flow of population into urban areas. Among the implemented measures was the abolition of the *nongzhuanfei*, a costly and restrictive fee that rural migrants were required to pay in order to convert their rural hukou to urban [16, 17, 37, 40].

In the late 1990s, these local hukou policies were consolidated, facilitating the migration and settlement of a greater number of individuals in urban areas on a more

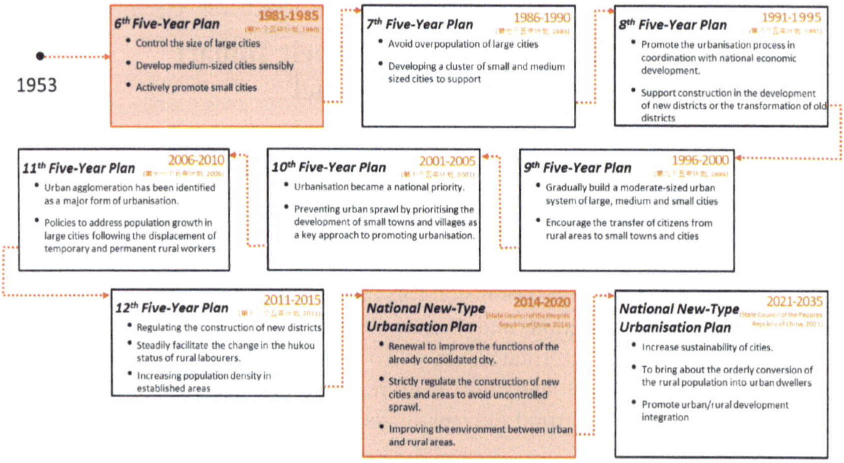

Fig. 6.1 Urbanisation plans and strategies 1981–2035 (*Source* Own elaboration)

stable basis. This, in turn, facilitated labour mobility and access to better economic opportunities, as well as promoting more inclusive and equitable economic growth by balancing development between China's rural and urban regions. However, it is evident that the system has continued to determine access to housing, education, and public services for Chinese citizens, despite the implementation of the aforementioned reforms [16, 17, 37, 40].

During this period, the hukou system constituted a pivotal element in China's urbanisation and social development policy, inextricably linked to the objectives delineated in the various Five-Year Plans since its inception in 1953. These plans have served as a guiding force in the evolution of the country's economic and social policy, including the way population mobility and residence are managed. Prior to the 10th Five-Year Plan (2001–2005), the hukou system underwent a transformation from a rigid mobility control mechanism to a more flexible instrument that facilitated Chinese urbanisation and balanced development. This edition represented a significant turning point, with three different novelties in terms of national urbanisation and the removal of institutional obstacles. First, it sought to promote the growth of smaller cities and facilitate the transition from rural to urban hukou, with the goal of enhancing the quality of life for migrants and supporting more inclusive and sustainable urban development [59]. Second, it emphasised the necessity of ensuring that new developments align with Chinese national conditions and prevent uncontrolled urban sprawl. Eventually, the development of small towns and villages was identified as a key approach to promoting urbanisation, with priority given to this strategy.

During the 11th Five-Year Plan (2006–2010), the focus was on enhancing the capacity of cities to accommodate the growing migrant population and implementing differentiated policies to manage urbanisation [65]. The Chinese government's approach involved the implementation of differentiated policies to guide urbanisation, which took into account the increase in population in large cities following the displacement of temporary and permanent rural workers [62].

The subsequent 12th Five-Year Plan (2011–2015) strengthened these policies for integrating rural workers into the urban system by modifying the hukou system to facilitate their transition to urban residents and improving their living conditions in small and medium-sized cities [43]. This Plan advocated for the synchronous organisation of industrialisation, urbanisation and agricultural modernisation, maintaining that industry should foster agriculture and cities should rely on rural areas to actively and steadily advance the urbanising process undertaken by the country. To achieve this objective, two kinds of measures were proposed. First, the establishment of a strategic pattern of urbanisation through the gradual formation of clusters of cities with a significant radiating effect. Second, changing the hukou status of rural workers to urban residents, while the population size and management in megacities was to be controlled and improved [5].

Over the past decade, China's urbanisation has attracted as much attention as its economic growth. While the process of urbanisation has undoubtedly promoted modernisation and wealth creation, it has also brought with it a number of problems and challenges, including the emergence of so-called 'ghost cities'. Consequently, in 2014, the Chinese government implemented a new urbanisation policy across the country, as evidenced by Chu [10] and Liu et al. [24]. In fact, the Chapter XV "Optimisation of urban spatial structure and management pattern new strategies" of the 2014–2020 National New-Type Urbanisation Plan set out the transformation of central urban areas and the construction of new cities and new districts [7]. One of the most significant issues is the Sect. 6.1 "Renewal to enhance central city functions", which encourages the relocation of specific existing functions in the central urban areas of megacities to satellite cities that support them, while reinforcing high-end services, modern commerce, innovation and other functions in the main urban cores of large and medium-sized cities. This is accompanied by the Sect. 6.2 "Strict regulation of the construction of new cities and areas", which sets out rigorous conditions for the creation of these new cities. The aim is to avoid uncontrolled expansion of urban limits. To achieve this, it is necessary to adopt an informed planning approach that takes into account factors such as population density, environmental capacity and available resources. Finally, Sect. 6.3, entitled "Improving the environment at the interface between urban and rural areas", addresses the improvement of the environment at the interface between urban and rural areas through the improvement of planning, construction, and management services in these areas. The objective is to promote balanced community development and to improve the living conditions of the rural population.

The current 2021–2035 National New-Type Urbanisation Plan considers urbanisation to be the sole path to modernisation. Its support is essential for the promotion of coordinated regional development [47]. Furthermore, the text acknowledges that while the Beijing-Tianjin-Hebei urban agglomerations have taken the lead in opening and development, the development of the central and western regions is relatively lagging. This implies that the spatial distribution and scale structure of cities and towns are not aligned with the carrying capacity of resources and the environment [52].

6.3 State-Led Interventions and the Challenges of Urbanisation

The relationship between the housing market and community interventions in the country is complex, multifaceted and intimately influenced by a series of public policies, economic demands and social changes that have developed over the last decades. The Chinese real estate market has experienced a massive expansion, characterised by rapid urban development and a remarkable growth in housing construction (Fig. 6.2). However, this has not been uniform, resulting in the emergence of different territorial challenges. These include housing affordability [25], regional imbalance [13, 63], the emergence of ghost cities [35, 38] and concerns about the quality of life in urban and rural communities [45, 46].

One of the most prevalent forms of intervention in urban contexts has been the enhancement of infrastructure and services within cities through the construction

Fig. 6.2 New urban developments around the centre of Taizhou city-prefecture in Jiangsu province (*Source* Photo taken by Javier Morán Uriel, 16 April 2023)

of new roads, the improvement of the public transport system, and the provision of essential services such as water and electricity supply. Such interventions not only enhance the quality of life for residents but also elevate the value of real estate in the affected areas [22]. In this context, the Chinese government has made a significant investment in upgrading urban infrastructure with the objective of supporting real estate development and encouraging housing demand in areas that previously lacked adequate services. This investment has been of particular significance in second- and third-tier cities, where the absence of infrastructure has constituted a significant impediment to economic growth and the attraction of new residents [49].

This phenomenon can be attributed to the fact that rural land is owned by agrarian collectives, while urban land is owned by the state. As China's urbanisation has expanded into rural areas, large tracts of farmland have been reclassified as urban by the local authorities themselves. This has resulted in the expropriation of land rights from the peasants who previously resided there, in exchange for an urban hukou, compensation, or a modern flat [57]. The strategy in question is based on municipalities purchasing rural land at a low price, subsequently rezoning it as urban and selling it at a high price to developers, thus generating a profit [27]. This encourages the current Chinese tax system to force local governments to rely on land sales, creating a situation where cities need to keep expanding and encouraging accelerated urbanisation of the territory [33].

In parallel, some cities have implemented policies of redevelopment and rehabilitation of older neighbourhoods seeking to renovate blighted areas and transform slums into modern and functional residential areas. This is achieved by improving the living conditions of existing residents and creating new opportunities for real estate development, all of which helped balancing housing supply and demand in cities [39, 44].

Another crucial aspect of the state-led interventions has been the implementation of affordable housing policies to address the needs of lower-income segments of the population. The Chinese government has implemented policies to facilitate the construction of low-cost housing and has established subsidy programmes to enable low-income individuals and rural migrants who have relocated to urban areas in search of improved prospects to access housing [42, 54]. These policies have been instrumental in addressing inequalities in access to housing and in guaranteeing that all citizens can live in decent conditions. Nevertheless, in first-tier cities, the discrepancy between the supply of and demand for housing has been a significant contributing factor to the surge in land prices, which has subsequently led to a dramatic increase in house prices [15, 61].

A final issue resulting from this type of state intervention has been an attempt to promote environmental sustainability within real estate development by adopting sustainable building practices, improving energy efficiency in buildings and creating green spaces and recreational areas in urban environments. As pointed out by Li and Mell [21], Li and Yao [20], Zhang et al. [58] and Zheng and Barker [60], these initiatives have been implemented through various projects that not only enhance the quality of life for residents, but also improve the desirability and value of properties in the affected areas in Maanshan and Beijing. The aforementioned projects illustrate the growing recognition of environmental sustainability as a mitigating factor in the context of significant environmental degradation and the loss of agricultural land, which is eroding local resilience [48, 50].

6.4 China's Density Distribution Challenges in Establishing Community Policies

In 2020, the People's Republic of China had a population density of 147.16 hab/km^2. When the populations of Macau, Hong Kong and Taiwan are included, this figure is reduced to 108.35 hab/km^2. The national surface reaches 9,593,693 km^2, but its population distribution is not uniform. In its vast and varied territory, two main regions can be distinguished: the eastern half and the western half. The former is watered by the Yangtze or Blue River and the Huang-Ho or Yellow River, has a high population density and different climatic conditions (continental, tropical and subtropical) that have traditionally favoured the settlement of the population. It is the economic base of the country due to its fertile cereal-growing areas of millet, sorghum, soybeans and wheat and its access to the sea. The western half of the country is mountainous and desert-like. The region is characterised by the Gobi and Taklamakan deserts and the Himalayas. The presence of these elements has a significant impact on the climate of this region, which has resulted in the formation of inhospitable and sparsely populated areas due to the hostile conditions they create.

6.4.1 The Heihe-Tengchong Line and the Territorial Inequalities It Represents

The significance of these zones and the aforementioned population stability is such that the imaginary Heihe-Tengchong Line is identified as a key element of the spatial distribution pattern of the population. This is characterised by a high population density in the east and a lower population density in the west, which still remains evident [23, 31, 51]. This pattern, characterised by flat terrain, low altitudes and proximity to coasts and rivers, has persisted over decades in the face of rapid economic and social development. Thus, a line can be drawn between the towns of Heihe in the northeast and Tengchong in the south, which would divide the nation into two distinct halves. This would result in a significant imbalance in terms of land area and population, with 57% of the land area comprising just 6% of the population, while the remaining 43% of the land area would have 94% of the population [34].

By considering the Heihe-Tengchong Line, three areas of high concentration of cities can be identified (Fig. 6.3). The first of these would be in the provinces of Guangdong and Guangxi on the southeast coast, with a privileged geographical position for maritime trade and connections with other Southeast Asian countries. The second zone is situated in the vicinity of Shanghai, which has emerged as the principal national financial and commercial hub, boasting a robust economy and a diverse range of employment opportunities, encompassing the financial and service sectors, as well as the manufacturing industry [9]. Finally, the third and final zone is the capital of the People's Republic of China itself. Beijing, the capital of China, has undergone significant investment in infrastructure, including roads, high-speed railways and public transport, which has facilitated the mobility of the population between Beijing and the neighbouring provinces of Hebei and Tianjin. This has led to a concentration of population in these areas.

6.4.2 The Evolution of Density in Three Main Administrative Levels

The uneven distribution of China's population represents a significant challenge for policy intervention (Fig. 6.4 and Table 6.1). In order to manage the population pressure that is present in densely populated regions, such as urban areas in the east, greater investment in infrastructure and social services is required. In contrast, less densely populated regions in the west have different needs, such as developing basic infrastructure and promoting social inclusion.

A comparison of population density by province reveals that the 847.91 hab/km^2 of Jiangsu Sheng and the 700.02 hab/km^2 of Guangdong Sheng stand out in comparison to the inland autonomous regions of Nèi Menggu Zìzhìqū with 20.05 hab/km^2 and Xīzàng Zìzhìqū with only 3.03 hab/km^2. A different case is that of municipalities under central jurisdiction, such as Shànghǎi Shì and Beijīng Shì (Fig. 6.5). These

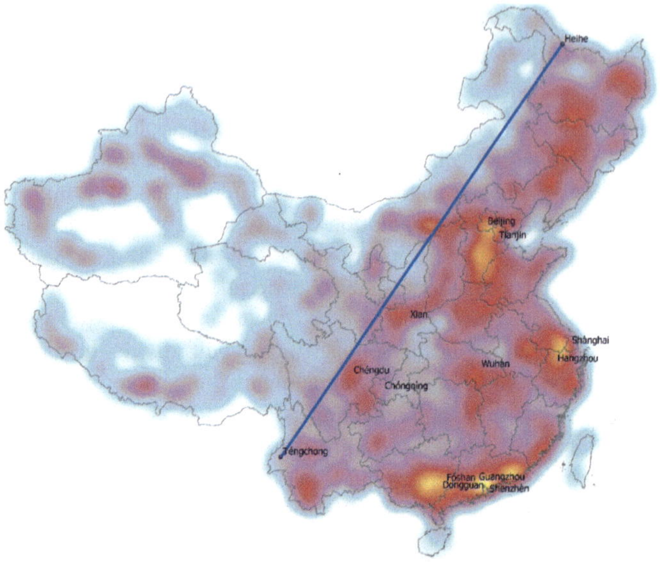

Fig. 6.3 The distribution of urban populations across the globe was analysed using the Kernel Density Method (*Source* Morán Uriel [28])

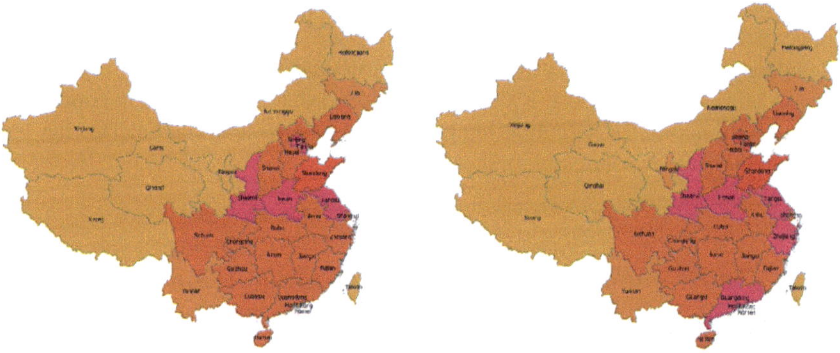

Fig. 6.4 The evolution of density in "Level 1. Province" between 2000 and 2020 (*Source* Own elaboration)

municipalities have a population density of 3922.24 hab/km^2 and 1334.05 hab/km^2, respectively.

The territorial discrepancies are maintained at the "Level 2. Prefecture". In this case, the differences between prefectures are once again strikingly evident. The city-level prefectures of Sānshā Shì, Ālāshàn Méng and Ālǐ Dìqu, which have lower population densities, are located in the western part of the region. There is only 0.01 hab/km^2 in the western part, and there has been practically no population variation

Table 6.1 Population distribution and density in China according to administrative Level 1 (*Source* Own elaboration based on data from China National Bureau of Statistics, Census and Statistics Department, SAR Hong Kong, Direcção dos Serviços de Estatística e Censos and National Statistics of Republic of China)

Denomination	Status	Surface (km²)	Population 2020 (Hab)	Density 2020 (Hab/Km²)
Ānhuī Shěng	Province	139,879	61,027,171	436.29
Běijīng Shì	Municipal Province	16,411	21,893,095	1334.05
Chóngqìng Shì	Municipal Province	82,403	32,054,159	388.99
Fújiàn Shěng	Province	123,756	41,540,086	335.66
Gānsù Shěng	Province	457,382	25,019,831	54.70
Guǎngdōng Shěng	Province	180,013	126,012,510	700.02
Guǎngxī Zhuàngzú Zìzhìqū	Autonomous Region	237,818	50,126,804	210.78
Guìzhōu Shěng	Province	176,140	38,562,148	218.93
Hǎinán Shěng	Province	34,259	10,081,232	294.27
Héběi Shěng	Province	189,809	74,610,235	393.08
Hēilóngjiāng Shěng	Province	472,766	31,850,088	67.37
Hénán Shěng	Province	165,467	99,365,519	600.52
Húběi Shěng	Province	185,776	57,752,557	310.87
Húnán Shěng	Province	211,842	66,444,864	313.65
Jiāngsū Shěng	Province	99,949	84,748,016	847.91
Jiāngxī Shěng	Province	166,939	45,188,635	270.69
Jílín Shěng	Province	190,282	24,073,453	126.51
Liáoníng Shěng	Province	147,076	42,591,407	289.59
Nèi měnggǔ Zìzhìqū	Autonomous Region	1,199,372	24,049,155	20.05
Níngxià Huízú Zìzhìqū	Autonomous Region	66,400	7,202,654	108.47
Qīnghǎi Shěng	Province	690,355	5,923,957	8.58
Shāndōng Shěng	Province	157,704	101,527,453	643.78
Shànghǎi Shì	Municipal Province	6,341	24,870,895	3922.24
Shānxī Shěng	Province	156,713	34,915,616	222.80
Shǎnxī Shěng	Province	205,624	39,528,999	192.24
Sìchuān Shěng	Province	484,056	83,674,866	172.86
Tiānjīn Shì	Municipal Province	11,610	13,866,009	1194.32
Xīnjiāng Wéiwú'ěr Zìzhìqū	Autonomous Region	1,644,707	25,852,345	15.72

(continued)

Table 6.1 (continued)

Denomination	Status	Surface (km^2)	Population 2020 (Hab)	Density 2020 (Hab/Km^2)
Xīzàng Zìzhìqū	Autonomous Region	1,204,776	3,648,100	3.03
Yúnnán Shěng	Province	383,195	47,209,277	123.20
Zhèjiāng Shěng	Province	104,873	64,567,588	615.67
People's Republic of China	Sovereign state	9,593,693	1,411,778,724	147.16
Macau	Special Administrative Region	3030	682,070	225.11
Hong Kong	Special Administrative Region	108,260	7,413,070	68.47
Taiwan	Republic of China	3,619,281	23,829,897	6.58
Total		13,324,264	1,443,703,761	108.35

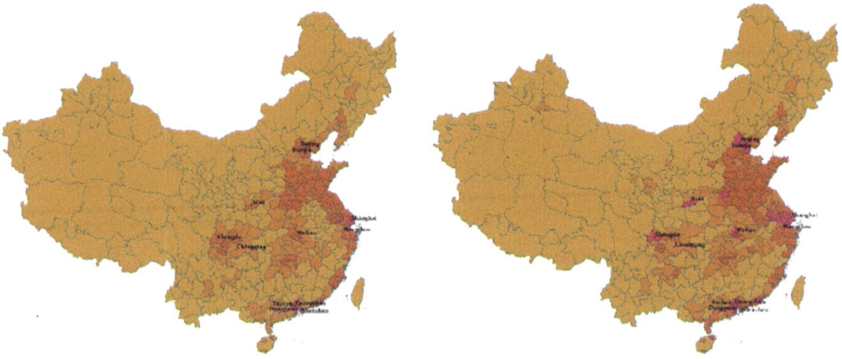

Fig. 6.5 The evolution of density in "Level 2. Prefecture" between 2000 and 2020 (*Source* Own elaboration)

in the period 2000–2020. The situation is entirely distinct from that of the city-level prefectures of Dōngguān Shì and Shànghǎi Shì, which have a population density of 4272.09 hab/km^2 and 3922.24 hab/km^2, respectively. Similarly, the situation differs from that of Shēnzhèn Shì, a sub-provincial city, which reached a population density of 8980.70 hab/km^2 in 2020. Moreover, the data indicates a significant variation in population density between 2000 and 2020. In particular, the city of Shenzhen has experienced an annual increase of 269.14 hab/km^2, while the other two cities have demonstrated a much more modest growth rate, with less than 100 hab/km^2 annually.

The population variations also demonstrate a clear tendency towards population concentration in the eastern part of the region, with the greatest density losses

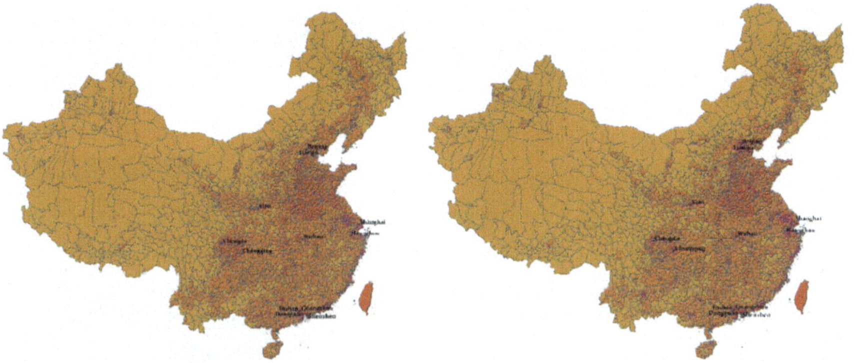

Fig. 6.6 The evolution of density in "Level 3. Counties" between 2000 and 2020 (*Source* Own elaboration)

observed in Nèijiang Shì and Ziyáng Shì Prefectures. In these areas, a decrease of approximately 10 hab/km^2 per year has been recorded, with values of − 9.47 hab/km^2 and − 8.50 hab/km^2, respectively. Conversely, Xiàmén Shí exhibits a variation of 97.89 hab/km^2, while Dongguan Shí displays 82.06 hab/km^2.

The total number of entities included in the "Level 3. Counties" is 2852, which comprises counties, county-cities, districts, and flags, both autonomous and not (Fig. 6.6). The variety of entities and the surface areas covered by each of them result in a significant difference between the maximum and minimum density values. Three density poles can be identified at this administrative scale, namely Beijing-Tianjin, Shanghai and Guangzhou-Dongguan, which are home to cities with a population of over nine million.

A first density pole includes the districts of Dongcheng, Fengtai, Haidian, Shijingshan, Xicheng, Chaoyang, Hebei, Hexi, Hongqiao, Jinnan, Nankai, Xiqing, Hedong, Heping and Tong Zhou districts collectively form a large conglomerate with densities all above 2000 hab/km^2. Other districts, including Changping, Daxing, Beichen, Dongli and Shunyi, also exhibit high densities, exceeding 1000 hab/km^2. A similar phenomenon can be observed in the second density pole, Shanghai, where a total of 10 districts collectively represents one of the most densely populated areas on the planet. In addition to the districts of Hongkou, Huangpu, Jiading, Jingan, Minhang, Songjiang, Yangpu, Changning, Putuo and Pudongxin, which have a population density of over 2000 hab/km^2, there are the districts of Fengxian, Jinshan and Qingpu, which have a lower population density. The third density pole comprises 24 entities covering 12,115.20 km^2. The 24 entities, which collectively cover 12,115.20 km^2, include Baiyun, Baoan, Dongguan, Futian, Guangming, Haizhu, Hong Kong, Huangpu, Jianghai, Liwan, Longgang, Longhua, Luohu, Nanhai, Nanshan, Panyu, Pengjiang, Pingshan, Shunde, Tianhe, Xiangzhou, Yantian, Yuexiu and Zhongshan. The combined population of these districts is 64,253,955, with a population density of 5303 hab/km^2 (Fig. 6.7).

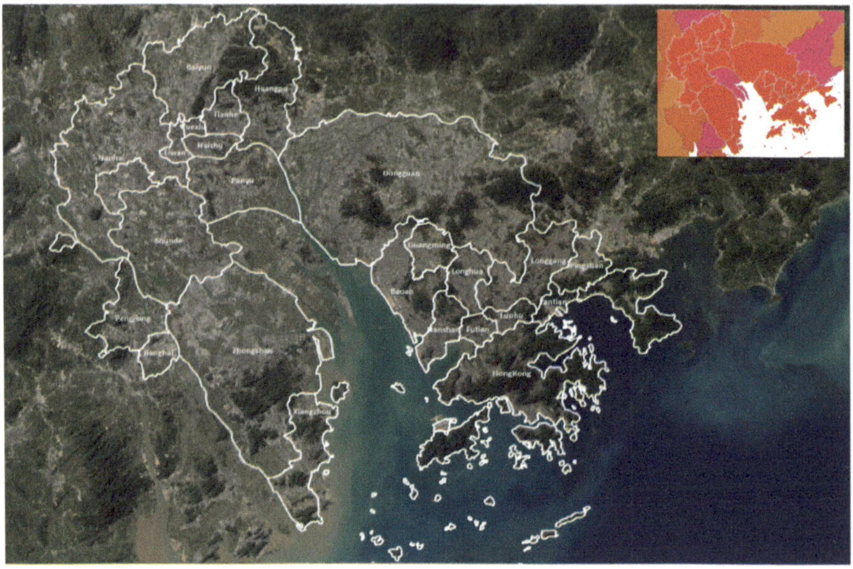

Fig. 6.7 Guangzou–Dongguan third density pole (*Source* Morán Uriel [28])

At the "Level 3. Counties", particularly dense entities can be identified. For instance, the Xinjiang Uyghur Autonomous Region, a western autonomous region, has densities close to and above 2000 hab/km^{2}in the districts of Shayibake, Shuimogou, Tianshan and Xinxi. In the latter district, the density reaches 3875.25 hab/km^2. In addition, the population density in Ruòqiāng Xiàn and Qiĕmò Xiàn counties is 0.22 hab/km^2 and 0.50 hab/km^2, respectively.

East–west division is reflected in the location of the 10 most densely populated areas. All of them have a population density above or very close to 25,000 hab/km^2 and are identified in the smallest of the administrative entities at this level, the districts. This density can be placed in context by noting that it is comparable to that of some cities with high population densities, such as Karachi (Pakistan) and Dhaka (Bangladesh). The former has a population density of approximately 46,657 hab/km^2, while the latter has a population density of approximately 30,900 hab/km^2, according to data from the latest report by Demographia [12]. Cities that not only experience congestion due to the high population numbers but also face significant challenges in terms of high demand for housing, which can result in the creation of substandard or prohibitively expensive housing, transport or public services, as well as a lack of availability of green space and recreational facilities [19, 26, 53]. In the Chinese case, the most dense districts (Table 6.2) are distributed across the provinces of Chóngqìng Shì, Guăngdōng Shĕng, Jiāngsū Shĕng, Shănxī Shĕng, Tiānjīn Shì and Zhèjiāng Shĕng. Furthermore, three of these districts are situated in Shanghai, one of the largest and most densely populated cities in the world. The high population density figures in this municipality directly under the central government are related to its importance as a national and international economic, financial, commercial

Table 6.2 Top 10 densest entities in China by administrative Level 3 (*Source* Own elaboration based on data from China National Bureau of Statistics, Census and Statistics Department, SAR Hong Kong, Direcção dos Serviços de Estatística e Censos and National Statistics of Republic of China)

Denomination	Province	Superface (km^2)	Population 2020 (Hab)	Density 2020 (Hab/Km^2)
Hépíng Qū	Tiānjīn Shì	10.01	355,000	35,455.37
Huángpǔ Qū	Shànghǎi Shì	20.35	662,030	32,528.89
Hóngkǒu Qū	Shànghǎi Shì	23.45	757,498	32,308.89
Bēilín Qū	Shǎnxī Shěng	22.99	756,840	32,922.66
Yuèxiù Qū	Guǎngdōng Shěng	33.67	1,038,643	30,845.82
Qiántáng Qū	Zhèjiāng Shěng	27.00	769,150	28,487.04
Liánhú Qū	Shǎnxī Shěng	38.40	1,019,102	26,536.07
Jìng'ān Qū	Shànghǎi Shì	37.00	975,707	26,370.46
Gūsū Qū	Jiāngsū Shěng	80.57	2,058,010	25,542.01
Yúzhōng Qū	Chóngqìng Shì	23.86	588,717	24,672.12

and maritime centre. Furthermore, the city of Shanghai is part of the fundamental framework of a scientific and technological innovation centre with a significant global influence. This is largely due to its close relationship with the Yangtze River Delta floodplain and its physical connection to Hangzhou Bay, which has made the port of Shanghai a major international maritime hub. This enables the city to connect the country to global markets and to handle the highest annual cargo volume compared to other major ports.

6.5 Discussion and Implications

China's vast land area and the remarkable disparity in population distribution present unique challenges in a country that is unlike any other. In densely populated areas such as the cities in the east, it is evident that greater investment in infrastructure and social services is required to effectively manage population pressure. Conversely, less densely populated regions in the west require policies that are geared towards basic development and the promotion of social inclusion. The unequal distribution of the population across China's vast territory implies that community intervention policies should be highly adaptable and tailored to the specific context of each region.

To date, the implemented policies have not resulted in a reversal of the Heihe-Tengchong Line, with only minor alterations in population density observed at the three scales analysed. Consequently, the existing inequalities in population distribution, which were intended to be balanced, have been maintained.

In addition, the variability in population density necessitates that community interventions should be tailored to the specific characteristics of the local area.

In urban areas, where high population density creates problems of congestion and demand for public services, policies should focus on improving the quality of life by creating green spaces and optimising public transport. In rural and less densely populated areas, however, interventions can be focused on sustainable development and improving access to essential services, promoting social cohesion and local economic development. This differentiated approach is of the utmost importance to guarantee that the policies implemented are both effective and aligned with the specific needs of each community, as suggested by studies that emphasise the integration of ecosystem assessments into land-use planning to protect natural resources and ensure sustainable growth [11].

One of the most significant challenges in densely populated regions is the scarcity of affordable housing and rising prices, which underscores the necessity for community-based initiatives that prioritize appropriate urban planning schemes. The emergence of 'ghost cities' in China prompts reflection on the desirability of specific times and densities in cities. This is particularly relevant in the context of cities with a population density below 5000 hab/km^2. It can be posited that community interventions must be based on accurate data and a thorough understanding of local needs to avoid creating unnecessary infrastructure in a context of energy and material crisis such as the current one.

Furthermore, the necessity for equitable development in China emphasises the significance of community intervention policies that facilitate equity in access to resources and opportunities, irrespective of population density. It is imperative that policies ensure that both densely populated, and less populated areas can benefit from economic and social development, thereby ensuring a fair distribution of resources and promoting sustainable development throughout the country. Meticulous planning and flexible implementation of community intervention policies are crucial to address regional disparities and promote balanced development in China.

The exponential growth of the global population over the past century, coupled with the rural exodus that has been occurring in recent decades, suggests that at some point the past will become too small to be inhabited and shared by people. In this context, a series of rapidly conceived and dehumanised cities or city peripheries have emerged. A phenomenon that originated in the so-called North American suburbs [1] and is currently at its greatest expression in the cities of Asia. This style of urbanism has resulted in the homogenisation of the peripheral areas of cities, transforming them into entirely monotonous and bland places [30].

These considerations open the floor for a further discussion on the urgency to assess the viability of maintaining identity in a country like China, which has a history spanning thousands of years, in an era of relentless construction and demolition, in a context where the urban population is expanding while the rural population is shrinking, as is currently observed. One of the notable consequences of the urbanisation process that has been encouraged by the state-led policies, in conjunction with the significant expansion of the construction industry, is the lack of identity, history and roots that is prevalent in most of the new urban areas. This has resulted in a vacuum of feeling and relationship between the inhabitants and their city. This situation is in part attributable to the persistence of outdated or inadequate building systems. In addition, a lack of sensitivity on the part of developers and local governments contributes to this problem. In many instances, the prioritisation of time efficiency over architectural quality has led to the utilisation of standardised designs with minimal complexity and variety. Consequently, the phenomenon of so-called 'superblocks' has become prevalent in the Chinese skyline, with these structures exhibiting a lack of meaningful connection to their socio-cultural environment. This results in a glaring disconnection with the surrounding urban landscape.

6.6 Concluding Remarks

Despite the evident advantages of community-based initiatives, several obstacles must be overcome to ensure that they have the greatest possible impact on China's real estate market. One of the principal challenges is regional inequality due to the concentration of significant proportion of investments and improvements in large first- and second-tier cities, while smaller cities and rural areas continue to face a lack of resources and opportunities. This unequal approach has resulted in an imbalanced distribution of development, with densely populated urban areas in the east undergoing a period of rapid urbanisation and economic growth, while the more rural and less densely populated western regions lag. The absence of investment in fundamental infrastructure and public services in these rural areas constrains the potential for development and perpetuates the country's economic and social disparity.

Furthermore, coordination between different levels of government and local communities is crucial to ensure that interventions are effective and aligned with the real needs of residents. The Chinese administrative structure, comprising multiple levels of government from central to local, presents a challenge to the implementation of coherent and effective policies. A lack of effective communication and collaboration between these levels can result in the duplication of efforts, a lack of resources in critical areas and the implementation of policies that do not reflect local realities. It is therefore of the utmost importance to reinforce community participation and ensure that the citizens' perspective is an integral component of the planning and implementation of development policies. Such measures not only enhance the efficacy and longevity of interventions, but also facilitate transparency and accountability in public management.

In the future, it is essential that property development policies and community interventions focus on the promotion of more balanced and sustainable growth across the country. Among these issues, the phenomenon of ghost cities stands out as a clear example of how initially unbalanced urban planning can, over time, achieve efficiency. It is therefore crucial that future policies are based on a thorough understanding of local dynamics and focus on the integrated development of urban and rural areas.

The creation of economic opportunities and the improvement of public services in these areas are of paramount importance if the country is to attract and retain residents and foster more equitable and sustainable development across the country. By concentrating policies on reducing the disparity between more and less developed regions, it is possible to achieve more inclusive growth that benefits a larger proportion of the population. This entails not only enhancing infrastructure and services in less developed regions, but also fostering local innovation and entrepreneurship, thereby creating an environment that enables communities to flourish in a sustainable and equitable manner.

Acknowledgements This work is the result of a collaborative effort between the authors. Federico Camerin has participated as co-author within the research project "*Las propiedades militares en Europa: Estrategias de Regeneración y sus efectos sobre Procesos de Producción Socio-Espaciales*" (Ramón y Cajal grant) funded by the Ministry of Science, Innovation and Universities, the State Research Agency through the State Plan for Scientific, Technical and Innovation Research 2021-2023, and co-funded by the ESF+.

CRediT Authorship Contribution Statement Roles played by contributors Javier Morán Uriel (JM), Federico Camerin (FC) and Rafael Córdoba Hernández (RC) to research outputs: Conceptualisation: RC; Data curation: JM and RC; Formal Analysis: JM and RC; Funding acquisition: FC; Investigation: JM; Methodology: JM and RC; Project administration: FC; Resources: FC; Software: JM; Supervision: RC; Validation: FC; Visualisation: JM and RC; Writing—original draft: JM, FC and RC; Writing—review and editing: JM, FC and RC.

References

1. Brenner N (2016) Critique of urbanization: selected essays. In: Critique of urbanization: selected essays. Birkhauser Verlag AG
2. Brewer B (2011) China urbanizes: consequences, strategies, and policies. Asia Pac Bus Rev 17(4):543–545. https://doi.org/10.1080/13602381.2010.481420
3. Brinkhoff T (2024) City population. https://www.citypopulation.de/
4. Bureaus of Statistics in China (2021) The people Bank of China: annual report. http://www.pbc.gov.cn/en/3688110/3688259/3689032/3709448/3984240/4850049/2023041215463143830.pdf
5. Casey J, Koleski K (2011) Backgrounder: China's 12th five-year plan
6. Census and Statistics Department SHK (2022) 2021 population census—main results. https://www.census2021.gov.hk/doc/pub/21c-main-results.pdf
7. Central Committee of the Communist Party of China (2016) The 13th five-year plan for economic and social development of the people's republic of china (2016–2020). https://en.ndrc.gov.cn/policies/202105/P020210527785800103339.pdf
8. Chan KW, Zhang L (1999) The hukou system and rural-urban migration in China: processes and changes. China Q 160:818–855. https://doi.org/10.1017/S0305741000001351
9. Cheng Y, Gao S, Li S, Zhang Y, Rosenberg M (2019) Understanding the spatial disparities and vulnerability of population aging in China. Asia Pacific Policy Stud 6(1). https://doi.org/10.1002/app5.267
10. Chu Y (2020) China's new urbanization plan: progress and structural constraints. Cities 103:102736. https://doi.org/10.1016/j.cities.2020.102736
11. Córdoba Hernández R, Camerin F (2024) The application of ecosystem assessments in land use planning: a case study for supporting decisions toward ecosystem protection. https://doi.org/10.1016/j.futures.2024.103399
12. Demographia (2023) Demographia world urban areas. In: Demographia, 19th annual edition, p 103. http://demographia.com/db-worldua.pdf

13. Fan S, Kanbur R, Zhang X (2011) China's regional disparities: experience and policy. Rev Dev Finance 1(1):47–56. https://doi.org/10.1016/j.rdf.2010.10.001
14. Gu S (2017) Transformation and healthy development of urbanization in China's 13th five-year national development period (2016–2020). China's Urban Socioecon Impact. https://doi.org/10.1007/978-981-10-4831-9_1
15. Gu H, Jie Y (2024) Escaping from "dream city"? Housing price, talent, and urban innovation in China. Habitat Int 145:103015. https://doi.org/10.1016/j.habitatint.2024.103015
16. Huang Y, Guo F, Tang Y (2010) Hukou status and social exclusion of rural–urban migrants in transitional China. J Asian Public Policy 3(2):172–185. https://doi.org/10.1080/17516234.2010.501160
17. Hui ECM, Yu KH, Ye Y (2014) Housing Preferences of Temporary Migrants in Urban China in the wake of Gradual *Hukou* reform: a case study of Shenzhen. Int J Urban Reg Res 38(4):1384–1398. https://doi.org/10.1111/j.1468-2427.2012.01176.x
18. Jiménez Moreno A (1991) Modelización cartográfica de densidades mediante estimadores Kernel. In: Treballs de la Societat Catalana de Geografia, pp 446–460
19. Kamble T, Bahadure S, Punglia S (2022) Availability and accessibility of urban green spaces in a high-density city: the case of Raipur, India. Prof Geogr 74(2). https://doi.org/10.1080/00330124.2021.2007495
20. Li B, Yao R (2012) Building energy efficiency for sustainable development in China: challenges and opportunities. Build Res Inf 40(4):417–431. https://doi.org/10.1080/09613218.2012.682419
21. Li Y, Mell I (2023) Green Infrastructure as Urban Melody: the integration of landscape principles into green infrastructure planning and design in China and the UK, pp. 117–132. https://doi.org/10.1007/978-3-031-18332-4_7
22. Li X, Che T, Li X, Wang L, Duan A, Shangguan D, Pan X, Fang M, Bao Q (2020) CASEarth poles: big data for the three poles. Bull Am Meteor Soc 101(9):E1475–E1491. https://doi.org/10.1175/BAMS-D-19-0280.1
23. Liu C, Wang F, Xu Y (2019) Habitation environment suitability and population density patterns in China: a regionalization approach. Growth Chang 50(1):184–200. https://doi.org/10.1111/grow.12283
24. Liu S, Jia L, Zhang F, Wang R, Liu X, Zou L, Tang X (2024) Do new urbanization policies promote sustainable urbanization? Evidence from China's urban agglomerations. Land 13(4):412. https://doi.org/10.3390/land13040412
25. Ma L, Liu Z (2024) How do local governments respond to central mandate in affordable housing policy? A qualitative comparative analysis of forty-one Chinese cities. J Urban Manage. https://doi.org/10.1016/j.jum.2024.05.003
26. Meng L, Zhu C, Wen KH (2020) Research on constructing a healing environment for the street spaces of a high-density city: Using street spaces in macao's old city area. Int J Environ Res Public Health 17(13). https://doi.org/10.3390/ijerph17134767
27. Mingye L (2017) Evolution of Chinese ghost cities. China Perspect 2017(1):69–78. https://doi.org/10.4000/chinaperspectives.7209
28. Morán Uriel J (2024) *Ghost cities en China: ¿distopía o predicción? Redefiniendo el concepto de estos lugares desde la densidad y el tiempo* [Universidad Politécnica de Madrid]. https://oa.upm.es/82561/
29. National Bureau of Statistics of China (2021) China National Bureau of Statistics. In: China statistical yearbook 2020. https://www.stats.gov.cn/sj/ndsj/2020/indexeh.htm
30. Pearson CA (1996) Asian cities: is 'generic' the wave of the future?: Rem Koolhaas stirs up strong emotions at a Singapore conference. Arch Rec
31. Qi W, Liu S, Zhao M, Liu Z (2016) China's different spatial patterns of population growth based on the "Hu Line." J Geog Sci 26(11):1611–1625. https://doi.org/10.1007/s11442-016-1347-3
32. Qingwu Z (1988) Basic facts on the household registration system. Chin Econ Stud 22(1):106
33. Rithmire ME (2017) Land institutions and Chinese political economy. Polit Soc 45(1):123–153. https://doi.org/10.1177/0032329216683167

34. Sang M, Jiang J, Huang X, Zhu F, Wang Q (2024) Spatial and temporal changes in population distribution and population projection at county level in China. Hum Social Sci Commun 11(1):288. https://doi.org/10.1057/s41599-024-02784-1
35. Shepard W (2015) In: Shepard W (ed) Ghost cities of China: the story of cities without people in the world's most populated country. Asian Arguments, Bloomsbury
36. Song Y (2016) Hukou-based labour market discrimination and ownership structure in urban China. Urban Studies 53(8):1657–1673. https://doi.org/10.1177/0042098015576861
37. Song Q, Smith JP (2021) The citizenship advantage in psychological well-being: an examination of the hukou system in China. Demography 58(1):165–189. https://doi.org/10.1215/00703370-8913024
38. Sorace C, Hurst W (2016) China's phantom urbanisation and the pathology of ghost cities. J Contemp Asia 46(2):304–322. https://doi.org/10.1080/00472336.2015.1115532
39. Sun Y, Li H, Lei S, Semple KT, Coulon F, Hu Q, Gao J, Guo G, Gu Q, Jones KC (2022) Redevelopment of urban brownfield sites in China: motivation, history, policies and improved management. Eco Environ Health 1(2):63–72. https://doi.org/10.1016/j.eehl.2022.04.005
40. Tao L, Wong FKW, Hui ECM (2014) Residential satisfaction of migrant workers in China: a case study of Shenzhen. Habitat Int 42:193–202. https://doi.org/10.1016/j.habitatint.2013.12.006
41. Tao L, Hui ECM, Wong FKW, Chen T (2015) Housing choices of migrant workers in China: beyond the hukou perspective. Habitat Int 49:474–483. https://doi.org/10.1016/j.habitatint.2015.06.018
42. Tian L, Yao Z, Fan C, Zhou L (2020) A systems approach to enabling affordable housing for migrants through upgrading Chengzhongcun: a case of Xiamen. Cities 105:102186. https://doi.org/10.1016/j.cities.2018.11.017
43. Wang B, Wu C, Kang L, Reniers G, Huang L (2018) Work safety in China's thirteenth five-year plan period (2016–2020): current status, new challenges and future tasks. Saf Sci 104:164–178. https://doi.org/10.1016/j.ssci.2018.01.012
44. Wang M, Zhang F, Wu F (2022) Governing urban redevelopment: a case study of Yongqingfang in Guangzhou, China. Cities 120:103420. https://doi.org/10.1016/j.cities.2021.103420
45. Wang D, Xie S, Wu J, Sun B (2023) The trend in quality of life of Chinese population: analysis based on population health surveys from 2008 to 2020. BMC Public Health 23(1):167. https://doi.org/10.1186/s12889-023-15075-2
46. Wang P, Li L, Wang Y (2023) The index construction and evaluation of quality of life in china and its spatial mismatch with subjective well-being: evidence from a city-level perspective. Soc Indic Res 170(3):955–985. https://doi.org/10.1007/s11205-023-03228-0
47. Wei H, Li L, Nian M (2021) China's urbanization strategy and policy during the 14th five-year plan period. Chin J Urban Environ Stud 09(01):2150002. https://doi.org/10.1142/S2345748121500020
48. Wei J, Chen Z, Kong X-Y, Zhang Y-J (2023) The prevention strategies for strengthening the resilience of urban high-rise and high-density built environment based on multi-objective optimization: An empirical study in Guangzhou, China. Environ Impact Assess Rev 101:107106. https://doi.org/10.1016/j.eiar.2023.107106
49. Wu J, Gyourko J, Deng Y (2016) Evaluating the risk of Chinese housing markets: what we know and what we need to know. China Econ Rev 39:91–114. https://doi.org/10.1016/j.chieco.2016.03.008
50. Xu C, Huo X, Hong Y, Yu C, de Jong M, Cheng B (2024) How urban greening policy affects urban ecological resilience: Quasi-natural experimental evidence from three megacity clusters in China. J Clean Prod 452:142233. https://doi.org/10.1016/j.jclepro.2024.142233
51. Yan N, Lin S, Chen J, Sun Y (2023) The power of culture: does diversity affect the breakthrough of the Hu Line in China? SAGE Open 13(3). https://doi.org/10.1177/21582440231194839
52. Ye Q, Qijiao S, Xiaofan Z, Shiyong Q, Lindsay T, Qin C, Changgui D, Daizong L, Tianle L, Jian L, Hong M, Ming Q, Ming T, Chengyue T, Ye W, Jin Z, Mengye Z (2020) China's new urbanisation opportunity: a vision for the 14th five-year plan. https://www.researchgate.net/publication/341188430

53. Yu P, Yung EHK, Chan EHW, Wang S, Chen Y, Chen Y (2023) Capturing open space fragmentation in high–density cities: towards sustainable open space planning. Appl Geogr 154. https://doi.org/10.1016/j.apgeog.2023.102927
54. Zenou Y (2012) Housing policies in China: issues and options. Reg Sci Policy Pract 4(4):393–419. https://doi.org/10.1111/j.1757-7802.2012.01083.x
55. Zhang H (2010) The hukou system's constraints on migrant workers' job mobility in Chinese cities. China Econ Rev 21(1):51–64. https://doi.org/10.1016/j.chieco.2009.10.002
56. Zhang L, Tao L (2012) Barriers to the acquisition of urban hukou in Chinese cities. Environ Plan A Econ Space 44(12):2883–2900. https://doi.org/10.1068/a4551
57. Zhang H, Li L, Chen T (2016) Where will China's real estate market go under the economy's new normal? Cities 55:42–48. https://doi.org/10.1016/j.cities.2016.03.014
58. Zhang F, Chung CKL, Yin Z (2020) Green infrastructure for China's new urbanisation: a case study of greenway development in Maanshan. Urban Stud 57(3):508–524. https://doi.org/10.1177/0042098018822965
59. Zheng L, Qunhui H, Tie L, Weifu Z (2007) The process and problems of industrialization and urbanization in China: the status of the tenth five-year plan and recommendations for the eleventh five-year plan. Chin Econ 40(1):6–30. https://doi.org/10.2753/CES1097-1475400101
60. Zheng W, Barker A (2021) Green infrastructure and urbanisation in suburban Beijing: an improved neighbourhood assessment framework. Habitat Int 117:102423. https://doi.org/10.1016/j.habitatint.2021.102423
61. Zheng X, Chen X, Yuan Z (2021) Exploring the spatial spillover effect of home purchase restrictions on residential land prices based on the difference-in-differences approach: evidence from 195 Chinese cities. Land Use Policy 102:105236. https://doi.org/10.1016/j.landusepol.2020.105236
62. Zheng J, He J, Shao X, Liu W (2022) The employment effects of environmental regulation: evidence from eleventh five-year plan in China. J Environ Manage 316:115197. https://doi.org/10.1016/j.jenvman.2022.115197
63. Zhou Z, Zhang A (2021) High-speed rail and industrial developments: Evidence from house prices and city-level GDP in China. Transp Res Part A Policy Pract 149:98–113. https://doi.org/10.1016/j.tra.2021.05.001
64. Zhou J, Hui EC (2022) The hukou system and selective internal migration in China. Pap Reg Sci 101(2):461–483. https://doi.org/10.1111/pirs.12651
65. Zhou N, Mcneil M, Levine M (2012) Assessment of building energy-saving policies and programs in China during the 11th Five-Year Plan. Energ Eff 5(1):51–64. https://doi.org/10.1007/s12053-011-9111-0

Chapter 7
Rereading the Role of Cafés as Urban Catalysts with a Focus on the Case of Tehran, Iran

Maryam Farash Khiabani and Mahsa Goldar

Abstract Regardless of where you are from, there is always a magic space that will celebrate what makes you, you. People gather there to be under the umbrella of somewhere away, in a simple café, to make connections, experience solitude, and gain the experience of being. From traditional coffee houses since the sixteenth century to contemporary cafés, they warmly open their doors to newcomers to their desires. This chapter aims to unravel the role of cafés as urban catalysts from their first establishment as traditional coffee houses in the Safavid era in Iran until now. The case study method among qualitative research designs is suggested to follow the transformation of coffee houses and cafés. Historical texts, travelogues, maps, figures, observations, spatial data, and Geographical Information Systems are used to make a spatiotemporal analysis of Tehran, the capital of Iran. According to Homi Bhabha's term, these hybrid spaces always hosted diverse customers. They were always on the border of being the hosts of diversity and homogeneity of different social classes. Magnificent coffee houses entertained Abbas the Great, the Safavid king, and his entourage. The humbler coffee houses, hosting lower-class guests, had a notorious reputation due to using young boys and opium. Coffee houses have played a significant role in shaping societal experiences over the centuries. During the Qajar period, they served as meeting places for specific guilds and even hosted artists. In the Pahlavi era, under the influence of intellectuals who had been to Europe, modern cafés appeared in the Iranian cultural sphere. After the Islamic revolution and a decade of war, their roles faded, but since 1990, new styles of cafés have emerged, leaving a lasting impact on Iranian everyday life. These spaces were always under the suspicion of the authorities and considered as places of deviation from ruling norms. Since the Safavid era, many Fatwas have been declared to restrict them. After a while, the governments could not resist the social forces and reopened them. This story has been continuing in Iran until now. More than a hundred cafés have been closed in recent months in Tehran due to political pressure, but the number of new ones is

M. Farash Khiabani (✉)
Department of Urbanism, University of Art, Tehran, Iran
e-mail: maryamkhiabani86@gmail.com

M. Goldar
Department of Urbanism, University of Science and Technology, Tehran, Iran

G. Siew et al. (eds.), *Diversity as Catalyst: Economic Growth and Urban Resilience in Global Cityscapes*, Urban Sustainability, https://doi.org/10.1007/978-981-97-8776-0_7

still rising. Furthermore, in other historical cities in Iran, urban regeneration projects are closely attached to the growth of café culture. The neglected part of society is now on the streets, presenting their lifestyle not only as a social hub but also as a strong business market. Consequently, we suggest reappraising cafés in cities. They are now centres of diversity and freedom, keeping society alive and resilient. In addition, reusing each building as a café gives it a new life. The revitalisation of the urban texture is indebted to these socio-commercial spaces. These all show that it is time to change our mindset and consider their functions beyond just entertaining spaces but mostly as diverse urban socioeconomic catalysts.

Keywords Coffee house · Café · Urban catalyst · Diversity · Other space · Hybridity

7.1 Introduction

No matter which city you are in, there will always be an attractive node to be free from the rest of the world. This host of diverse groups has its own unwritten laws. The importance of coffee houses' role in the cities' social-cultural-physical structure since their inception is evident in history by the abundance of their mentions in many travelogues. For instance, in Çelebi's descriptions of the cities he passed through, he counted the number of mosques, baths and shops, and the number of coffee houses ([15], p. 466). In Iran as well, the number of these spaces was counted; Isfahan had 19 coffee houses in the early Qajar period, and Tehran had 2334 coffee houses in the late Pahlavi period ([5], p. 289; [43], p. 45). On the way visitors travelled, they mentioned coffee houses as a sign to reach the settlements [1].

The presence of diverse social classes in these coffee houses was one of the reasons for alluring the people. The attraction of coffee houses caused grave suspicion among governments and traditional classes of societies in Iran and other countries. Being different and non-compliant with existing structures was considered a deviation over the years. Nevertheless, the world's perspectives have changed. According to Sandercook [46], cities now face various social, economic, and even gender groups, colours, and races; therefore, a discussion of cultural diversity was created. In response to her question of why difference is seen as a problem, she considers the values and norms of the dominant culture to be the reason for this and points out the effects of this discrimination on urban planning principles. Her new question is, "*How can 'we,' all strangers to each other, live together in the city?*". The understanding of diverse cultures has led experts to believe that different groups are looking for various and diverse spaces ([14], p. 14). Some unfettered places are trying to liberate us from the predicaments of old beliefs and have holistic impacts on the various aspects of cities, from cultural to economic [55].

The experience of traditional Iranian coffee houses compared to modern cafés elucidates the strong connection of diverse activities to these spaces. This research studies this connection by reviewing the historical background of coffee houses, their transition into trendy cafés and their role as generators, especially in making the old parts of the city attractive in the socio-political urban sphere of Iran, with emphasis on the case of Tehran. This process is also investigated by focusing on the spatial transformation of these spaces over time.

7.2 Coffee and Coffee House History (from World Scale to Iran)

During coffee expansion, it first spread over the areas close to the main cultivation lands in Ethiopia (Abyssinia) and Yemen [3, 51]. Coffee is also called Islamic wine ([22], p. 27). In the sixteenth century, it reached Hijaz, Egypt, and the Levant and had significant cultural and social messages [30]. The need to consume this drink in a group influenced the formation of a particular place [53]. When coffee was brought from Yemen to Hejaz in the sixteenth century, coffee houses spread to places in Mecca and other Islamic cities worldwide ([13], p. 346). The Ottoman Empire and the Kingdom of Iran also open the door to the new drink. This plant probably entered Iran from Egypt and Turkey ([39], p. 275). According to Chardin ([16], p. 277) Safavid coffee houses had the most free-thinking atmosphere possible. Poets and missionaries stood in the middle of the coffee house, reciting the *Shahnameh* or religious texts.

Coffee houses were among the Islamic world's first and most important non-religious socio-cultural institutions ([18], p. 57), where strangers become brothers. These places allowed men to mix and participate in safe political, economic, and cultural discourse, even in Europe in the eighteenth century. Habermas calls this public realm neither a new social system nor a high utopia but a social hub representing the political, civil, economic and moral bourgeoisie ([31], pp. 105–106). Meanwhile, coffee houses were known for corruption in the cities under the rule of the Ottoman Empire, from Jerusalem to Damascus, because of their freedom and the formation of political and forbidden discussions, preventing believers from worshipping, etc. [53]. At first, in 1570, there were concerns about emptying mosques due to coffee houses in Islamic cities ([50], p. 20).

7.3 Café, the Host of Hybridity and Diversity

Cafés are spaces that bring together individuals with diverse backgrounds and experiences. These spaces encourage individuals to express their opinions and perspectives, allowing differences to be acknowledged and respected. According to Homi Bhaba, socio-cultural spaces like cafés deal with identity, hybridity and liminal spaces. Originally, hybridity refers to individuals who are born to parents of different racial backgrounds [9]. These hybrid spaces are ambivalent and liminal ([35], p. 30). They are social spaces experienced in everyday life and expanded to create a hybrid culture and identity that unifies the cohesion of time, place, and people ([42], p. 64).

This hybridity made coffee houses a 'third space' where diverse discourses and contradictory identities coexist harmoniously. This space is free from binary limitations and thinking, and it has several advantages in creating an atmosphere of interaction and dialogue ([9], p. 37). This cultural interaction may result in two undamaged cultures [32]. It is where diverse cultural groups unite to share their multifaceted memories ([35], p. 16). Meanwhile, culture's importance is linked to the economy, development, industry, and, in summary, all aspects of today's societies. According to Sen [47], the topic of culture and its relationship with development, in response to the question of "*Is culture only a complement to development?*" explains a decisive perspective: It is the process of increasing one's freedom to find one's goals. He considers the adoption of cultures as the determining factor of social and economic progress. This is the point that is evident in the transformation and spatial expansion of Tehran's cafés in recent years [23]. The connection of a different lifestyle, which is joint with the café subculture and the robust economy, caused the explosion of these spaces.

This issue, the combination of social, cultural, and economic aspects in one place, has been one of the main characteristics of coffee houses since their inception and seemingly continues until now in Iran. Reviewing the literature and historical travelogues helps recognise their role in previous eras. It can be followed by studying new forms of these places in contemporary Tehran, the capital of Iran. Moreover, apart from cultural issues, their apparent effect on the city's physical, economic, and social aspects encourages us to consider cafés from a different lens as an urban catalyst.

7.4 Lessons from Urban Catalysts

Since the 1970s, the term 'catalyst' has been used in urban texts and projects to improve urban areas and developments [8, 44]. Cambridge Dictionary Cambridge [21] defines catalyst as: 'someone or something that causes change or a substance that makes a chemical reaction happen more quickly'. In the first descriptions, elements that can boost urbanisation and spatial transformation in a city and beyond scales can act as urban catalysts. They can be physical and non-physical features like an event, a work of art, or an urban artefact and take precious value from their activities ([44], p. 87). Some researchers define urban catalysts as a means of revitalisation or regeneration of the urban area [11, 20, 49, 52]. It can be a hotel, a shopping mall or a cultural venue ([6], p. 45). Also, some strategies can be proposed for sustainable and prosperous developments ([20], p. 19; [27], p. 19; [40]).

In the case of urban catalysts, some other terms, such as 'urban flagship' in the first steps and 'urban magnet' later on, have been used. Flagship projects, such as the Bilbao Museum, attracted the attention of researchers trying to estimate the impacts of these mega projects on the context and found the consequences controversial [28, 36]. Urban magnet is another term that has recently referred to different kinds of catalysts, such as creating a space that sparks curiosity and invites transformation, which is possible by designing for a specific subculture. Contrary to popular belief, making places appealing to a broad range of people is sometimes designed for no one; the target groups and subcultures must be considered [29]. Table 7.1 reviews some significant texts and views about urban catalysts.

One of the common issues in most of these projects was using them as a means of urban revitalisation or a kind of generator, which must be Identifiable [6, 11, 49]. Recent approaches have evident the role of subcultures, community needs, and moving beyond physical features to be liveable rather than just a magnificent artefact. Meanwhile, it is noticeable that most of these projects are mega projects, and the value of small spaces seems underestimated. Since a wide range of creative classes live in the cities due to the increased numbers of academics, artists, researchers, and designers and the fact that art, culture and creativity are no longer dominated by rulers [26], some new phenomena have emerged (Fig. 7.1). In the subsequent paragraphs, cafés, considering their strong background in the history of Tehran and the cultural-artistic places surrounding them, are studied as 'urban catalysts'.

7.5 Methodology

A qualitative design was used in this research, and the 'case study method' was chosen based on Creswell [19] to explain the role of 'cafés' and 'coffee houses'. As a method, case study has a long history and has been widely used in politics, social studies, management, planning, etc. This method is mainly based on the Chicago School of Sociology ([10], pp. 276–277). Data collection in this method is generally extensive and includes several sources of information such as documents, observations, interviews, and audio-visual materials. In this study, the reviewed documents were historical travelogues, books, articles, maps, etc. Observations, interviews, and some spatial data were collected to study the case of cafés in today's Tehran.

The next step involved a detailed analysis in which the researchers addressed several major issues [19]. During the Pahlavi and pre-Pahlavi periods, information about our case, coffee house, and café was extracted based on documentary research, texts, and figures. Their approximate locations were used as point features in ArcMap to make a temporal map. Then, using density functions, we generated an approximate density distribution map within the Tehran city boundaries during the Pahlavi period.

To compile information related to cafés in Tehran in 2024, after conducting observations and some interviews, we used the QGIS software program and retrieved relevant layers from OpenStreetMap (OSM) under the 'amenity' subset. All cafés in Tehran were selected as point features and checked using our observation data. Then, we generated a spatial distribution map for the cafés using ArcMap. In the subsequent step, we used OSM data to create separate point layers for other cultural spaces, which we considered to play catalytic roles along with cafés, such as art centres, theatres, museums, and bookstores; our previous studies in the central zone of Tehran were also contributed [23, 24]. Finally, by overlaying the generated raster

Table 7.1 Urban Catalysts' background in the literature, by authors adapted from [6, 8, 11, 20, 27–29, 40, 44, 49, 52]

Scholar/ researcher	Date	Title	Views on urban catalysts
Rossi	1970, 1982	The architecture of the city	• Physical and non-physical feature • An urban artefact
Wejchert	1974	Elements of urban composition)	• The role of creativity as a probable urban catalyst
Attoe and Logan	1989	American urban architecture: catalysts in the design of cities	1. The new element modifies the elements around it 2. Existing elements are enhanced or transformed 3. The catalytic reaction does not damage its context 4. The understanding of the context 5. All catalytic reactions are not the same 6. Catalytic design is strategic 7. A product better than the sum of the ingredients 8. The catalyst can remain identifiable
Walkowiak and Frazier	2000	Brownfields redevelopment as a catalyst for creating sustainable cities	• The opportunity to increase sustainability • The important role of community needs • As a means of regeneration and economic growth
Sternberg	2002	What makes buildings catalytic? How cultural facilities can be designed to spur surrounding development	• An activity generator • An anchor • An amenity that promotes development
Bohannon	2004	The urban catalyst concept	• A as a means for revitalisation • A strong sense of place • Shaped along with the context
Grodach	2008	Museums as urban catalysts: the role of urban design in flagship cultural development	• Flagship projects as a means of economic development • The need for compatible activities • Emphasis on existing arts-based activities
Davis	2009	Urban catalysts in theory and practice	• As a means of urban renewal • Engendering social activities • Diverse uses • Sustainability in social, economic and spatial aspects
Oswalt et al.	2013	Urban catalyst: strategies for temporary use	• Temporary uses as urban catalyst • The importance of the role of available resources • Emphasis on the shared values of the community • As a strategy toward sustainable development

(continued)

Table 7.1 (continued)

Scholar/ researcher	Date	Title	Views on urban catalysts
Francin	2015	Urban catalyst	• Creating a distinct image by becoming a symbol within the context • As an asset to the city • The necessity of a holistic and adaptable process • The vital role of community needs
Balvočienė and Zaleckis	2021	Cultural urban catalysts as meaning of the City	• As the meaning of the city • City as a network • The importance of cultural nodes • Identity of the place
Haden, Holland and Irvine	2020	Urban magnet	Designing for a specific activity subculture, authenticity, identity, and art are essential Six elements of attractor places" essential to creating Urban Magnets 1. The activity subculture that makes the place unique 2. What speciality retail supports the activity? 3. Is there space to support production related to the subculture? 4. Is there a place to learn about it? 5. Programming and events to showcase the activity 6. Unique urban form that brings it to life

layers, we produced a density distribution map for cafés, art centres, theatres, and bookstores to show the relation between these spaces.

7.6 Tehran City and Its Coffee Houses and Cafés' Transformations

Tehran, the capital of Iran, has a population of around 9 million. In 1786, Agha Mohammad Khan, the first king of the Qajar dynasty, chose this city as the capital of Iran, which has remained since then. Before, Isfahan, the capital of the Safavid era, was the country's most magnificent city. Various studies have been done on the function of Iran's modern cafés, especially in Tehran and other big cities [7, 23, 24, 37, 38]. These spaces underwent many transformations over the years, which we will review below.

Fig. 7.1 A row of shops along with Henry Corbin St. in the centre of Tehran, including the Vast Gallery and platform, café, art and handicraft shop, from their official Instagram account in 2024

7.6.1 From the Safavid to the Qajar Era

In Safavid Iran, coffee houses were connected to religious circles and located near religious schools. Some Religious officials objected to immoral customs such as using young boys and smoking for entertainment. Moreover, the Safavid rulers' disapproval of coffee houses was evident even in classifying these spaces besides brothels and notorious jobs such as dancing and singing. For instance, the *Mash'al-Dar-bashi*, an official who collected their taxes, supervised coffee houses ([34], pp. 238–239).

With the emergence of coffee houses, the minstrels who used to present their art in the streets and squares found a special place and contributed to the prosperity of the coffee houses ([12], p. 86). During the Qajar era, coffee houses were the origin of 'Coffee House Arts' such as painting. It was a place where the tradespeople gathered after their daily work and spent hours listening to religious and *Shahnameh's* stories. The painters visualised the *Naqali Majles* by drawing the stories on the walls of the coffee houses. These artefacts attracted a larger audience and boosted the business ([54], p. 14).

During this era, we faced the issue of changing the pattern of public consumption from coffee to tea, and various factors played a role in this change.[1] The main drink served in coffee houses also changed from coffee to tea. However, the atmosphere of

[1] For more information, the Tea chapter in the Encyclopaedia of Iranica is available: https://www.iranicaonline.org/articles/cay-tea.

the coffee houses remained the same, and despite this change, they were still known as coffee houses ([4, 34], p. 321). At the end of this era, the role of some coffee houses,[2] changed from entertainment hubs to gathering places for people attending political events, especially during Tehran's constitutional revolution [45].

7.6.2 The Pahlavi Era

The Pahlavi era was the beginning of trendy cafés in Iran. After the arrival of radio in Iran in 1940, people were more eager to hear the news, especially during World War II, than to listen to stories and poems, which made the storytellers gradually disappear from the scene. The popularity of television after 1958 caused more severe damage to the institution of the coffee house. In addition, the influence of Western culture made the former customers prefer to spend their free time in modern cafés [4].

Iran started to follow the global modernisation process, influencing its urban areas. New urban streets were designed, and for the first time in Iranian history, urban planning projects ran to transform Iran from a traditional society to a modern one [33]. Some elites who visited Western countries returned to Iran and looked for new places for contemporary needs (Fig. 7.2) [2, 41]. During this time, modern cafés hosted intellectuals and their meetings, whereas the traditional coffee houses did not serve coffee anymore due to the expansion of tea as the conventional Iranian drink; they remained the gathering place of traditional classes, especially men. Modernisation allowed women to be present in cafés, so they were no longer masculine spaces [23, 48].

During this period, some cafés in the city centre were social hubs for poets, writers and artists. *Jalal Al-e-Ahmad*[3] spoke harshly against the government in *Firouz Café*, and many young people came to listen to his speeches. *Khosrow Golesorkhi*[4] and *Ahmad Shamlou*[5] also visited this cafe. These activities raised the government's suspicions [2].

The separation of coffee houses and cafés began in the Pahlavi era. The map in Fig. 7.3 shows all their approximate locations and the expansion of these spaces from the beginning of the Qajar era (1789) in the area of the Grand Bazar and Naseri's wall[6] to the end of the Pahlavi period (1979) around Tehran's modern streets, such as Jomhuri (known as Shah St.), Enghelab (known as Reza Shah St.), and Keshavars (known as Elizabeth Blvd.), located in the north part of the Tehran's old

[2] A coffee house stood until around 1950 in the current Baharestan Square on the side of the parliament. Later on in the Pahlavi era, on the day Dr Mohammad Mossadegh was taken from his house to the parliament when he refused to attend a parliamentary meeting, the military forces fired into the crowd, and people sheltered Mossadegh inside this coffee house [45].

[3] Writer, social and political critic born in 1923 and died 1969.

[4] Journalist, poet and communist activist born in 1944 and executed in 1974.

[5] Poet, journalist and political activist born in 1925 and died in 2000.

[6] Neighbourhoods located into the wall of Tehran during the Qajar king, Naseredin Shah.

Fig. 7.2 Loghanteh café in the central district of Tehran in the Pahlavi era (captured by authors from Loghanteh Café Museum's archive in Tehran)

city. Nevertheless, due to the lack of spatial data, it is impossible to authenticate the distribution of coffee houses; that is why approximate zones were added to the map.

7.6.3 From the Islamic Revolution Until Now (2024)

After the Islamic Revolution of 1979, due to the political changes and the Iran-Iraq war, cafés were insignificant places, or at least there were very few documents about them in those years. Regarding religious laws, mingling between boys and girls was not allowed, which was one of the main concerns about coffee shops at that time, but they survived through all problematic circumstances. In the past three decades, cafés' transformations have been evident, but restrictions remained. They flourished during the 1990s and the following decade in Tehran. From the mid-2010s until now, another wave of cafés has been formed, using them as public arenas. Today, cafés have become a part of the culture, mainly developed mostly by young people and host diverse groups with different expectations ([23, 25], p. 43). Some customers see these spaces as reminiscences of old Iranian houses and culture in vernacular architecture, including the yard, comfort foods, cutleries, etc. Others use them as social workspaces or study rooms in the modern context (Fig. 7.4).

Cafés, regardless of their designs and customers, are spreading in Tehran's urban areas. Figure 7.5 illustrates this spatial expansion on the map. Although cafés can be

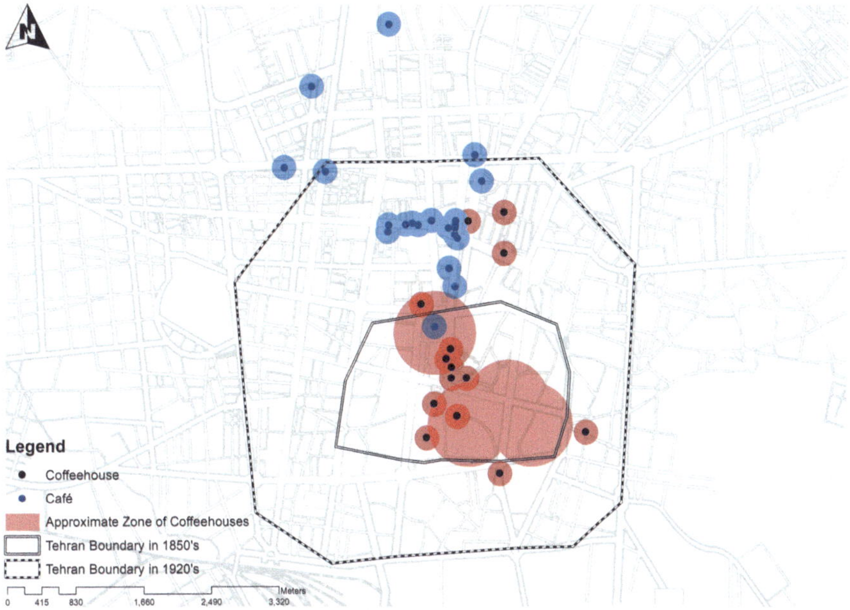

Fig. 7.3 Temporal map of coffee houses and cafés from Qajar to the Pahlavi era (by The Authors)

Fig. 7.4 Reusing old buildings as a café in Tehran, a modern style at the left and the traditional one with a courtyard at the right (captured by authors)

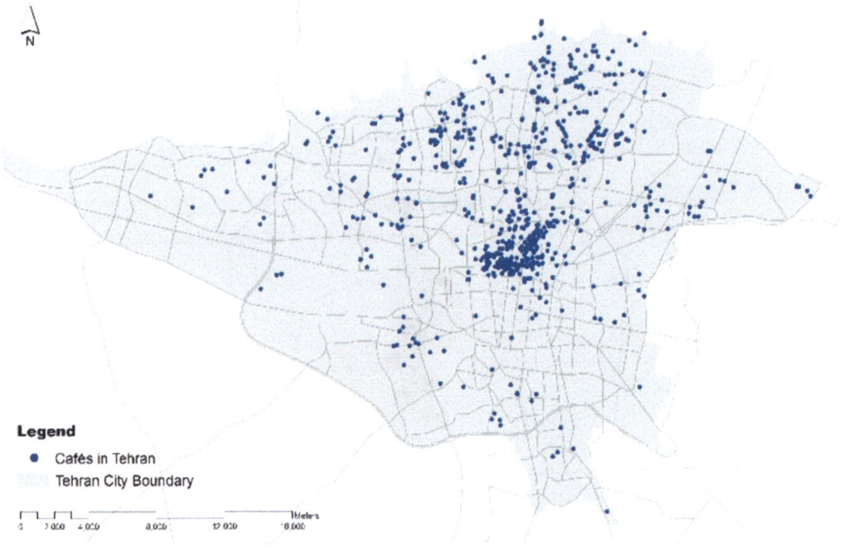

Fig. 7.5 The expansion of the cafés in Tehran in 2024 (by The Authors)

seen all over the city, they are mostly in the central district. The northern areas are second in accommodating them.

In addition, after some political changes in recent years, the issue of hijab caused many restrictions in these spaces, and many of them were closed due to disobeying the dress code rules by their customers. The news about their closures is primarily available on their official Instagram accounts. It seems that after around 400 years, cafés are still under suspicion by the Iranian government. This suspicion seems to have caused the overlooking of cafés' significant capacities as social hangouts and even economic boosters and their influential role in the urban space. Their economic impacts aligned with their social and cultural roles in the socio-economic structure, making them considered urban catalysts.

7.7 Discussion

Today, the expansion of café culture is evident in contemporary Tehran. Needless to say, the quality and richness of life and spatial-social interactions depend on diversity [17]. Iranian cafés and their different users, especially those in traditional houses, work on restoring their lost sense of identity by making a welcoming place for diverse groups. Limiting the community based on government norms or traditions is unnecessary. This diversity can even lead to renovating old buildings and urban areas, which has less cost for the governments if they take supportive and advocating

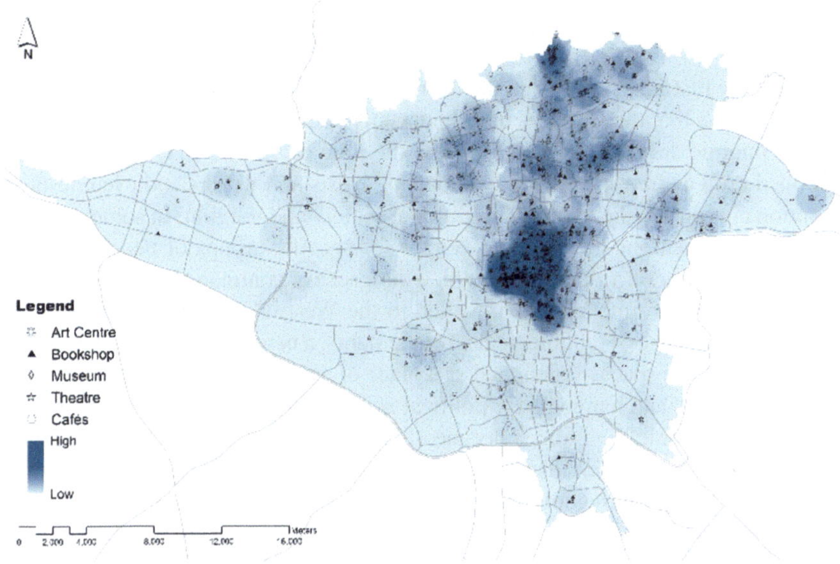

Fig. 7.6 The map of the distribution of cultural and artistic places and cafés in 2024 (by The Authors)

roles to allow these spaces to fulfil their socio-economic capacities actively instead of directly intervening. The 'private' sector runs and expands cafés, but their impacts are huge on a 'public' scale. They can be divided into cultural, social, perceptional, economic, and physical aspects. Each café is a single and small project that gives space for a breath of fresh air, and its accumulation creates an area with a different atmosphere, using the power of art and culture to revitalise the urban context. The map of the distribution of cultural and artistic places and cafés reveals the vigorous transformations and connections between these places in Tehran (Fig. 7.6).

Today, the burgeoning presence of cafés in Tehran is intricately intertwined with cultural spaces. Cafés play a vital role in developing these spaces; in return, they frequently pave the way for new cafés in the vicinity. Consequently, they collectively foster the growth of a vibrant subculture within the city. The café, a potential social place for local communities and one of the tourist attractions in Iran, is unfortunately neglected by the government. However, these spaces, loved by their users and the private sector, function as urban catalysts. They can transform our society if we view them and their capacities differently. The bottleneck to optimise the cafés' capacities has always been ideological and dominant rulers in Iran, from the Safavids' suspicions of coffee houses to the issue of hijab and gathering boys and girls in the contemporary era.

7.8 Conclusion

Reviewing the historical texts from the formation of the old coffee houses to present-day cafés shows that these spaces have constantly been gathering places for various social, economic and cultural groups. From the time of Safavid to Qajar, they were mentioned both as free spaces and hangouts of artists and writers and even nests of corruption contrary to the values of traditional society and religion. In some periods, coffee houses functioned as meeting venues for specific groups of people, similar to the headquarters of a particular guild. This process continued until the Pahlavi era when the Iranian intellectuals travelled to the West; along with the worldwide socio-political changes, technological advancements, a new form of cafés emerged in Iran. Some of these spaces functioned as a place of entertainment but others created intellectual spaces and hangouts for political activities.

During the Islamic Republic, a decade after the war, Iran entered a new political era when dialogue among civilisations and freedom of expression became new trends, which resulted in more freedom in the public sphere. The expansion of cafés in the central district of Tehran, especially around the University of Tehran, along with the line of Enghelab St. bookstores and cultural spots, marked the beginning of a new cultural movement. As stamping grounds of the traditional class, coffee houses concurrently stayed unchanged as men-only.

Since the 2010s, the café has become one of the most essential and up-to-date symbols of Iranian modernity, sometimes with a solid connection to the past identity, with the help of new social networks such as Instagram. From the interior to the exterior design, the details used in cafés, e.g. furniture, cutlery, cuisine, etc., make an essential contribution to the business and play a crucial role in customer attraction. The Digital Revolution has removed limitations and boundaries and helped cafés no longer be exclusive to students or a particular social class; they also welcome family gatherings, diverse ages and genders.

From this period, we see the spread of cafés in the more prosperous areas of Tehran, such as the northern district and mainly in the city centre. Furthermore, they influenced the renovation of the urban fabric by reusing old buildings. The activities associated with cafés connected the business to social life, e.g. galleries, bookstores, cinemas, theatres and art shops. Nowadays, imagining these spaces without a café is less possible. The role of the creative class is very prominent herein.

It is noteworthy that despite all the social, spatial, and economic impacts of cafés in Tehran, especially in the last three decades, the Iranian government is still wary of them. The city is in flux, but cafés are considered deviant and should be under surveillance. This research suggests that Iran's urban planning policies must acknowledge these spaces' prominent socio-economic and cultural role in the contemporary urban context. For social stability, we need to hear other voices and welcome different spaces that host diverse groups. With the help of multi-disciplinary views in culture, sociology, economics, architecture, urban planning, art, etc., programmes should be formulated to use these spaces as urban catalysts.

Acknowledgements This article did not receive any financial support. We appreciate the owners and users of studied cafés who keep them alive as an oasis of hope.

References

1. Afzal-Ul-Molk GH (1982) Travelling to Khorasan and Kerman. Toos (in Persian)
2. Akhavan Langeroodi M (2011) From Naderi Café to Firooz Café. Morvarid (in Persian)
3. Al-Arnaout M (2012) From the cultural history of coffee and cafés. Jadawel (in Arabic)
4. Al-e-Dawoud SA (1992) In: Yarshater E (ed) Coffee in the encyclopedia Iranica, vol V. Routledge & Kegan Paul
5. Astrabadi MMK (1998) In: Anvar A (ed) Jahangosha-ye Naderi. Association of Cultural Works and Honors (in Persian)
6. Attoe W, Logan D (1989) American urban architecture: catalysts in the design of cities. University of California Press. http://ark.cdlib.org/ark:/13030/ft5k4006v5/
7. Azad Armaki T, Shalchi V (2005) Two worlds in Iran: Mosque and coffee shop. Ir Assoc Cult Stud Commun 1(4):163–183 (in Persian)
8. Balvočienė V, Zaleckis K (2021) Cultural urban catalysts as meaning of the city. Arch Urban Plan 17(1):16–28. https://doi.org/10.2478/aup-2021-0002
9. Bhabha HK (1994) The location of culture. Routledge
10. Blaikie NWH (2000) Designing social research: the logic of anticipation polity
11. Bohannon CL (2004) The urban catalyst concept
12. Bolukbashi A (1996) Iran's coffee houses. Cultural Research Office (in Persian)
13. Burckhardt JL (1819) Travels in Arabia. African Association for Promoting the Discovery of the Interior Parts of Africa
14. Carmona M, Wunderlich F (2012) Capital spaces: the multiple complex public spaces of a global city. Routledge. https://doi.org/10.4324/9780203118856
15. Çelebi E (1999) The Journey to Egypt, Sudan and Abyssinia, vols 1 and 2 (H. M. Masri, Trans.). Dar Al-Afaq Al-Arabiya Publishing (in Arabic)
16. Chardin J (1957) Chardin's Voyages: travels in Persia (M. Abbasi, Trans.). Amirkabir (in Persian)
17. Cheshmehzangi A, Li HMA (2020) Innovation through urban diversity and achieving comprehensive sustainable urbanism from a community-oriented approach. Curr Urban Stud 8(2):222–240. https://doi.org/10.4236/cus.2020.82012
18. Çobanoğlu Ö (2009) Cultural interrelationships between Turkish minstrel tradition and Egyptian folk culture in the socio-cultural context of coffeehouses in Alexandria and Cairo. Int J Mod Anthropol 2:57–65
19. Creswell JW (2007) Qualitative inquiry and research design: choosing among five approaches Sage Publication
20. Davis J (2009) Urban catalysts in theory and practice. Urbanism 13(¾):295–306
21. Dictionary C (2024) Cambridge
22. Ellis M (2004) The coffee house, a cultural history. Phoenix
23. Farash khiabani M (2021) Unravelling the essence of otherspaces in the city of Tehran (central district) with emphasis on the cultural context (University of Art). Tehran (in Persian)
24. Farash Khiabani M, Shahabian P (2019) Learning from cafés: cafés as other spaces. Motaleate Shahri 8(30):109–121. https://doi.org/10.34785/J011.2019.407 (in Persian)
25. Fazeli N (2011) Coffee shop and city life. Tisa (in Persian)
26. Florida R (2012) The rise of the creative class, revisited. Basic Books
27. Francin K (2015) Urban catalyst. School of Urban Planning. McGill University, Research Project

28. Grodach C (2008) Museums as urban catalysts: the role of urban design in flagship cultural development. J Urban Des 13(2):195–212
29. Haden B, Holland M, Irvine B (2020) Urban magnets: how activity subcultures can be a catalyst for rejuvenating cities. Independently Published
30. Hattox RS (1996) Coffee and coffee houses: the origins of a social beverage in the medieval near east, 3rd edn. University of Washington Press
31. Hetherington K (1991) The Badlands of modernity: heterotopia and social ordering. Routledge. https://doi.org/10.4324/9780203428870
32. Lin J, Mele C (eds) (2012) The urban sociology reader, 2nd edn. Routledge. https://doi.org/10.4324/9780203103333
33. Mashayekhi A (2019) The 1968 Tehran master plan and the politics of planning development in Iran (1945–1979). Plan Perspect 34:849–876. https://doi.org/10.1080/02665433.2018.1468805
34. Matthee RP (2005) The pursuit of pleasure: drugs and stimulants in Iranian history. Princeton University Press
35. McLaughlin M (2013) Crossing cultural borders—a journey towards understanding and celebration in aboriginal Australian and non-aboriginal Australian contexts. Curtin University, Australia
36. Meades J (2017) The Bilbao effect. In: The Spectator. https://www.spectator.co.uk/article/the-bilbao-effect/
37. Mehdizadeh SM, Azizi F (2016) The situational analysis of communicative action in Tehran cafes. Cultural Stud Commun 11(41):69–100. https://www.jcsc.ir/article_19085_29512b5ec5e31eab3794cdb2a48a13e6.pdf (in Persian)
38. Moeini SHI, Arefian M, Kashani B, Abbasi G (2018) Urban culture in tehran urban processes in unofficial cultural spaces. Springer
39. Olearius A (2018) The travels of Olearius in seventeenth-century Persia (J. Davies, Trans.; L. Jenott, ed.). Forgotten Books Publisher
40. Oswalt P, Aksamija A, Reijndorp A, Rötzer F, Sassen S (2013) Urban catalyst: strategies for temporary use
41. Pooladvand M (2011) Naderi cafe: the history of cafes and cafes in Iran. Manoochehri (in Persian)
42. Prasetyo F (2012) Sidewalk warfare: run and learn new curatorial constellation. Japan Foundation
43. Ravandi M (1975) Social history of Iran. Amir Kabir Publications (in Persian)
44. Rossi A (1982) The architecture of the city. The MIT Press
45. Salour K (1995) Ein al-Saltanah Diary, vols 1 and 3. Asatir Publication (in Persian)
46. Sandercook L (2000) When strangers become neighbours: managing cities of difference. Plan Theory Pract 1(1):13–30
47. Sen A (2001) Development as freedom, 2nd edn. Oxford University Press
48. Shahri J (1992) Old Tehran, vols 1 and 2. Moen Publication (in Persian)
49. Sternberg E (2002) What makes buildings catalytic? How cultural facilities can be designed to spur surrounding development. J Arch Plan Res 34–42
50. Ukers WH (1922) All about coffee. In: The tea and coffee trade journal company
51. Van Arendonk C (1913) 'Ḳahwa' in the encyclopaedia of Islam (T. W. A. M. Th. Houtsma, R. Basset, R. Hartmann, ed.). Brill
52. Walkowiak E, Frazier D (2000) Brownfields redevelopment as a catalyst for creating sustainable cities. WIT Trans Ecol Environ 39:113–122. https://doi.org/10.2495/URS000131
53. Yildiz MC (2010) Coffeehouses as informal education institutions and coffeehouses in Egypt. Proc Social Behav Sci 9:1362–1367

54. Zikrali N (2014) Coffee house painting: a national and religious culture display. Nedaye Tarikh Institute and Publications (in Persian)
55. Zukin S (1995) The cultures of cities. Basil Blackwell, Oxford. https://asarartmagazine.ir/19173/artist/%D9%85%D8%B9%D8%B1%D9%81%DB%8C-%D9%88-%D8%A2%D8%AB%D8%A7%D8%B1-%D8%AD%D8%B3%DB%8C%D9%86-%D9%82%D9%88%D9%84%D9%84%D8%B1-%D8%A2%D9%82%D8%A7%D8%B3%DB%8C/. Accessed 2 Apr 2024

Chapter 8
Brownfields in China: The Case of Liaocheng Chemical Plant

Armin Mehdipour and Jon Kellett

Abstract In recent decades, sustainable brownfield regeneration has been adopted as a pragmatic policy approach to pursue the ideal of efficient use of natural resources and to contain the far-reaching implications of urban sprawl in many nation states. Since the 1990s, regeneration of urban brownfields has become an integral part of city policy and action in the developed world. In other places, such as China, brownfield regeneration is still emerging. Currently, China lacks a comprehensive legislative and regulatory framework for addressing brownfield sites. Nonetheless, the Chinese central government is initiating a nationwide brownfield redevelopment framework through pilot studies on a smaller scale. This study focuses on a specific small-scale industrial brownfield in eastern China, the Liaocheng Chemical Plant, which has been extensively utilized for decades. This case exemplifies China's efforts to decommission outdated, polluting industries and rehabilitate land through environmental remediation measures. This research distinguishes itself by examining brownfield land regeneration within the broader context of urban and structural changes in China's regional and political landscapes. Through a detailed case study analysis, incorporating personal interviews and site visits, this study sheds light on the practical application of China's evolving policy and regulatory framework for preventing and managing land contamination.

Keywords Urban Change · Economic change · Brownfields · Policy and practice · China · Liaocheng chemical plant

8.1 Introduction

Over the last century, cities around the world have grown in both population and physical extent. The rapid process of urbanization and spatial redistribution of population have occurred at different times and in different ways in different parts of the world. Concurrent with expansion of urban areas has been the outward migration of

A. Mehdipour (✉) · J. Kellett
School of Architecture and Civil Engineering, University of Adelaide, Adelaide, SA, Australia
e-mail: armin.mehdipour@adelaide.edu.au

G. Siew et al. (eds.), *Diversity as Catalyst: Economic Growth and Urban Resilience in Global Cityscapes*, Urban Sustainability, https://doi.org/10.1007/978-981-97-8776-0_8

industrial activities away from city cores towards outer suburbs, regional areas and overseas [2, 31, 32]. The relocation or closure of factories are commonly driven by factors such as constraints on the supply of land in inner-city areas, the increasing value of urban land and more importantly structural change resulting from international competition [16]. One result of this structural shift is a legacy of vacant and abandoned sites in inner urban locations, namely brownfields. Currently, there are diverse interpretations of the term "brownfield" worldwide. Initially, the United States Environmental Protection Agency [35] defined brownfields as abandoned, idled, or under-used industrial and commercial sites with real or perceived environmental contamination. However, several European countries such as the UK, Germany, and France have adopted a broader definition, labeling brownfields as Previously Developed Land (PDL), encompassing both contaminated and non-contaminated derelict land [14, 24, 30]. In the light of this variation in definitions, for the purpose of this research, brownfield is discussed as "previously developed land with former industrial activity that is currently unoccupied or not in productive use and is known or suspected to be contaminated". As a result of long-term and extensive industrial activity, brownfields are often found to contain hazardous materials in the soil and groundwater which pose a serious threat to human health and ecological system.

The rapid growth of urban areas arising from the economic and geopolitical trends of the last century has resulted in an increasing volume of brownfields within cities across the developed and developing world. The substantial expansion of urban areas, along with the relocation of industrial activities, stands as the principal driver behind the emergence of brownfields [25]. Generally, as cities undergo uncontrolled growth due to rapid urbanization pressures, inner-city sites with a history of manufacturing activity often remain unused and abandoned. This scenario is prevalent in the formation of urban brownfields in many developed and developing countries, including China. Disproportionate urban land growth emerged as a predominant trend across China from the early 1980s to the early 1990s, particularly in major coastal cities where economic growth surged rapidly [7, 33]. As suggested by national statistics, during the 10-year period between 1985 and 1995, the urbanized area in China increased by 486% [41]. Much of the existing literature [11, 12, 15, 36, 37, 40, 42, 43] have attributed China's urban growth to the post-Mao economic growth, marketization of urban land and institutional/political forces. The Chinese model of urban development is strongly connected with the economic restructuring, and government jurisdictions over land reform and land-use change. Since the introduction of Chinese economic reform in the late 1970s and unprecedented economic growth of the country, the economic value of land has been increasingly recognized by government officials in China [41]. High land revenue has prompted local governments to increase land sales, solidifying it as the primary source of local finance in recent decades and shaping urban growth policies at the local government level [23]. With the implementation of a market economy, several land-related reforms have been adopted by the Chinese government which substantially increased the municipal power and pressure in marketization of land. This has been considerably influential in acceleration of urban expansion in China.

Similar to the process of urban land expansion, the relocation of industrial enterprises from central urban to suburban and rural areas has been predominantly influenced by the acceleration of the China's economic and urban land reforms. As the reforms progressed and cities grew at a rapid pace, the concentration of old manufacturing sites in urban cores has been recognized piecemeal as a serious economic and environmental problem [4, 13]. Therefore, many industries located in central cities, especially in traditional industrial-based cities such as Beijing and Shanghai, closed due to market forces for land relocation. Under the economic consideration, high land prices resulting from urban growth act as a strong stimulus to the relocation of industrial land, particularly from the coastal cities to inland cities [3]. This has been mainly driven by growing foreign direct investment in coastal land development promoted by local governments and their designated enterprises. On the other hand, many traditional manufacturing industries located in inner cities have been dealing with serious environmental degradation and pollution issues. Under both economic and environmental considerations, local governments in China have been more responsive to the challenge of industrial relocation.

China's economic reforms and urban land policies have spurred local governments to make substantial efforts in implementing brownfield regeneration policies on a large scale. Nevertheless, there is currently no comprehensive legislative model or regulatory framework in place for brownfields in China [26]. The Chinese central government plans to introduce the brownfield redevelopment framework initially on a smaller scale through pilot studies, aiming to establish a robust basis for assessing its potential for broader implementation and full coverage in the longer term [34]. This study aims to examine one of these small-scale brownfield sites in China that has been intensively used by industrial activity for decades. Located in a small town in eastern China, the Liaocheng Chemical Plant serves as an example illustrating the Chinese government's efforts to decommission old, polluting industries and repurpose the land following necessary environmental remediation measures. In light of this case study analysis, the chapter aims to examine how the Chinese policy and regulatory framework for the prevention and treatment of land contamination has been applied in a practical case.

8.2 Research Methods

This study utilizes both secondary and primary data. Secondary data include government reports, policy directives, and academic literature. These secondary resources are systematically identified for review based on their comprehensiveness, credibility, and relevance to the subject matter. The immaturity of brownfield policy in China necessitated the collection of primary data in addition to the secondary sources. Primary data were collected through interviews and intensive field surveys of a specific site case study. The interview participants were city officials, professionals, public development agencies, academics, and urban researchers. Overall, 12 semi-structured face-to-face interviews were undertaken in three Chinese cities, including Beijing, Jinan and Liaocheng, between 2017 and 2020. The emphasis of the interviews was on identifying the traditional policy drivers and hierarchy of planning approaches to brownfield regeneration in light of the political and socio-economic structure of China. The study employed a 'Snowball Sampling Method' (also known as chain sampling, chain-referral sampling or referral sampling) for identifying interview participants and case study sites. These research team at Shandong University, who were experts in the field and well-connected with industry and academia, guided us to specific individuals for both interviews and visits to the case study site. Respondents have been anonymized in the reporting of interview data. Given the semi-structured nature of the interviews, only data relevant to addressing the research objectives have been included in this study.

The remainder of this chapter is composed of three sections. Section 8.3 introduces the former site of the Liaocheng Chemical Plant. The purpose of this section is to provide a general overview of the site, in terms of its geographical location, former use and environmental condition. Section 8.4 underlines possible stimuli to the redevelopment and reuse of this brownfield site. Having elaborated on different aspects of the site redevelopment, it first discusses the existing government directives and action plans associated with the relocation of polluting industries and remediation of contaminated soil. It then explains how the clean-up process has been implemented at the site. Finally, Sect. 8.5 offers a comprehensive insight into the broader picture of brownfield issues in China, discussing the overall state of urban-industrial structure and examining the relationship between Chinese brownfield policy and practice.

8.3 Case Study Overview: Liaocheng Chemical Plant

Covering a total area of around 15 ha, Liaocheng Chemical Plant is located 1 km northwest of a suburban town called Liangshui. The plant and the town are connected by Changsheng Road (Fig. 8.1). Liangshui has a total population of 67,000 people and a total area of 142 km^2, comprising 108 small villages [8]. Liangshui is an agricultural town, and its economy is predominantly based on agricultural production. However, there are several industries and factory sites led by private and national enterprises that are scattered around the town, such as Chang Yun Edible Fungus Company, Jinhong Food Company, and Tai Lin Chemical Co. Ltd. The most notable and the largest enterprise of the town was Liaocheng Chemical plant which ceased its operation in 2013. The growing industrial activities and, particularly, the rapid development of private enterprises have been influential in keeping the economic engine of the town and surrounding cities running.

Liangshui is located in the northwest of 'Liaocheng City' in Shandong province in China. As of the end of 2016, the total area and total population of Liaocheng city were recorded to be 8715 km^2 and around 6 million inhabitants [18]. Liaocheng is administratively divided into eight counties, including one district (Dongchangfu), one county-level city (Linqin), and six county-level districts (Yanggu, Dong'e, Chiping, Gaotang, Guan and Shen) [18]. Liaocheng Chemical Plant is located in

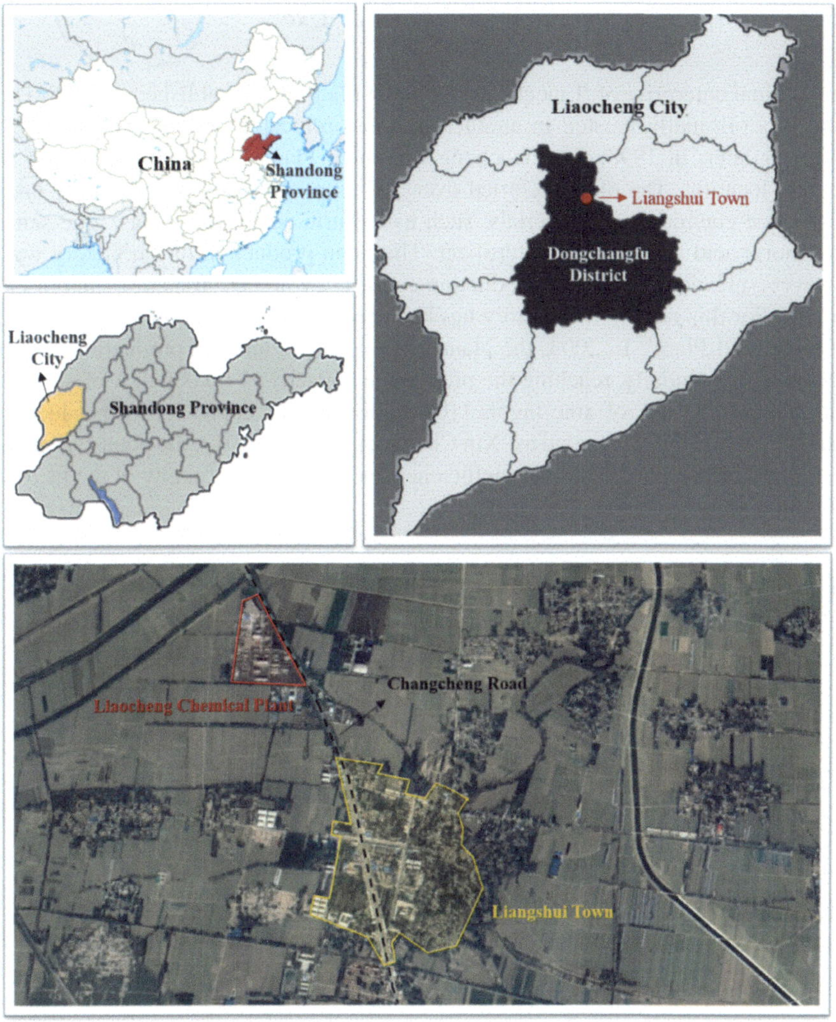

Fig. 8.1 Regional overview of the Liaocheng chemical plant

Dongchangfu district, which covers a total area of 829 km^2 and includes a total population of 899,000 people [8]. Figure 8.1 shows the geographical location and regional overview of the plant in the county, city, state and country level.

8.3.1 Production Overview and Former Use

The original enterprise of 'Liaocheng Chemical Plant' was established in 1975. The plant was originally located in another area and relocated to Liangshui in 1997. One year later, in 1998, the Liaocheng Chemical Plant started its production in the new location. The plant recorded over 500 employees and was formerly used to produce chemical raw materials, such as sulfuric acid, zinc ingots, zinc sand, phosphoric acid and phosphate fertilizer. The main production of the factory was, however, sulfuric acid, with a production capacity of 30,000 Tons per Annum (T/A) [10]. For this reason, this factory has been commonly referred to as 'Liaocheng Sulfuric Acid Plant'. In 2003, the plant was expanded, and the production output increased substantially, reaching the production capacity of 60,000 T/A of sulfuric acid and 30,000 T/A of zinc ingots [10]. Following the expansion of production volume, the plant was renamed as 'Xin Chemical Plant'. In 2013, the plant ceased its overall production due to market inefficiency and economic difficulty of the factory (Fig. 8.2).

Fig. 8.2 Dereliction of the former Liaocheng chemical plant (*Source* photos taken by the authors in 2017)

8.3.2 Soil Contamination Issue

After the closure of the Liaocheng Chemical Plant in 2013 and following intensive monitoring, the soil and groundwater at the site were found to be highly contaminated (see Fig. 8.2). As a result of long-term and extensive chemical activity of former plant from the production to the final disposal stage, the soil and groundwater became widely polluted by hazardous materials and toxic heavy metals, such as zinc, arsenic, lead, cadmium and copper. According to the detailed environmental analysis conducted by the Environmental Protection and Planning Institute [10], the soil and groundwater contamination problems at the former site of the Liaocheng Chemical Plant were mainly caused by four principal factors, including:

- The long-term production of raw materials and product stacking,
- The process of transportation of materials,
- Ineffective disposal of hazardous materials and toxic heavy metals (given the long-term production history of the site, a great amount of industrial waste scattered and buried on and under the soil)
- The lack of environmental protection facilities and technologies which could prevent or control the spread of contaminates.

Having analyzed the main sources of pollution at Liaocheng chemical plant, EPPI's survey identifies four key factors in the spread of contaminants over the soil surfaces and groundwater system across the site, including:

- Horizontal migration of contaminants at the surface of the soil (through runoff, sedimentation, stacking and dust diffusion) which resulted in absorption of pollutants into wider soil surfaces in the migration path.
- Vertical or downward migration of contaminants (mainly through the rainfall) which resulted in pollution of groundwater and lower strata of the soil.
- Horizontal or vertical migration of contaminates as the groundwater flowed through the lower surfaces of the soil.
- Atmospheric diffusion of pollutants caused by flue gas and wind which spread out the contaminants over a large surface area of the site.

8.4 Possible Stimuli to the Site Redevelopment

Based on information gathered from the field-survey and personal interviews, two main factors have predominantly stimulated the redevelopment of Liaocheng site, including the government's attempt to control industrial pollution as well as the steady growth of Liangshui city. These two stimuli are briefly discussed in the following sub-sections.

8.4.1 Government Attempts to Control Industrial Pollution

The first stimulus to the closure and redevelopment of the former Liaocheng Chemical Plant is associated with environmental issues and severe pollution of the site. Since the late 1980s, the national government has made several attempts to prevent and control environmental pollution sources caused by urban industries all over China [25]. In 1989, the government formulated the first environmental act as 'Environmental Protection Law of the People's Republic of China'. This was a general law aiming at improvement of the living environment, protection of public health, and promotion of sustainable economic and social development. Following the act of 1989, there have been several relevant national laws, directives and legal notices addressed by the government to control the environmental pollution problems caused by various industrial activities. The notable examples of these national environmental laws and notice are as follows:

- "Opinions on Strengthening Soil Pollution Prevention and Control"; introduced in 2008 to establish a fundamental management system framework for soil pollution prevention and control at the national level.
- "The Protection of Industrial Enterprises to Redevelop the Use of Environmental Safety Notice"; as a general law promulgated in 2012. This Act provided nine guiding principles on the protection and remediation of the soil and groundwater contamination caused by heavily polluting enterprises that have been already shut down due to bankruptcy or relocation.
- "Soil Pollution Prevention and Control Action Plan" (Commonly known as Soil 10); that was introduced in 2016 to provide some guidelines and basic support for the promotion of soil and groundwater pollution prevention in such industrial sites where the nature of the original land-use has changed.

Apart from above-mentioned environmental action plans and legal notices, the Chinese government has addressed several directives that strongly encourage the closure, relocation and environmental treatment of old and polluting industries. In 2004, an important directive was issued by the Ministry of Environmental Protection, namely "Notice on the Work of Preventing and Treating Environmental Pollution in the Process of Enterprise Relocation". This was a very general directive aimed at limiting and controlling environmental incidents caused by soil pollution on already relocated or relocating industrial sites, at three levels of site investigation, clean-up, implementation and waste disposal [28]. In 2014, the Ministry of Environmental Protection issued another national directive on relocation of polluting industries and soil contamination issues, namely "Notice on Strengthening Pollution Prevention and Control of Industrial Enterprises in Restraining, Relocating, Redevelopment and Utilization of Site". The central objective of this legal directive was to enforce relocating or closed industrial enterprises to promptly announce the quality of soil and groundwater of the site prior to and within the closure process. Meanwhile, according to this directive, local environmental protection departments at all levels should urge the relocation of polluting industrial enterprises, following the process of public pollution prevention [29].

8.4.2 *Steady Growth of Urban Area*

One main issue that has significantly stimulated the redevelopment of the former Liaocheng factory site is connected to the steady growth of the urban area. As cities grow, large-scale factories (such as Liaocheng Chemical plant) that were once located outside become surrounded by urban development which prompts cessation of operation. Their sites provide land for redevelopment. This has been a common scenario for many Chinese cities since economic reforms started in the late 1970s and early 1980s. As indicated by the Chinese government's official report, the number of cities and towns increased from 2911 in 1981 to 17,632 in 1995—an absolute increase of over 500%—[41]. This was most significant for small towns around the cities, such as Liangshui. The number of towns in China was recorded to be 16,992 in 1995, 6.34 times of that in 1981 [41]. After the mid-1990s, disproportionate expansion of urban land in China continued to become an obvious urbanization trend. As statistics suggest, the total area of urbanized land in China increased from 3.4 × 104 km^2 in 1990 to 6.0 × 104 km^2 in 2010, an increase of over 75% [17]. In the coastal zone (where Liaocheng City is located), this figure has been the greatest, from 1.7 × 104 km^2 in 1990 to 3.2 × 104 km^2 in 2010, a substantial increase of over 88%.

It is important to recognize the key role of the local government in substantial expansion of urban land in China. In the case of Liaocheng city, the municipal government has been actively involved in promoting the development of the city area over the past two decades or so. In 2001, after the approval of the Shandong Provincial government, the Liaocheng City Planning Bureau presented a comprehensive development vision, so called 'Liaocheng City Master Plan (2001–2020)'. The central objective of this master plan was to determine new layouts for land-use planning and development of Liaocheng City, within three layers of core area, Dongchangfu district and Liaocheng city [19]. Meanwhile, having defined various development zonings and new urban areas, the municipal government introduced regulations to support the implementation and management of the master plan. In October 2014, the master plan was revised and reformatted as 'Liaocheng City Master Plan (2013–2030)' (see Fig. 8.3). According to the master plan, by 2030, the urbanization level of Liaocheng city will reach 68%, with an anticipated urban population of 4.68 million people (of which 1.8 million people are in the central city) [19]. More importantly, as proposed by the master plan, to ensure the healthy, orderly and harmonious development of economy and society in Liaocheng City, it was deemed necessary to define new layouts for spatial organization of major industries within the designated zonings. For this reason, having addressed critical land-use planning and development issues, the Liaocheng City government's master plan has provided a strong stimulus to the closure and redevelopment of former 'Liaocheng Chemical Plant'.

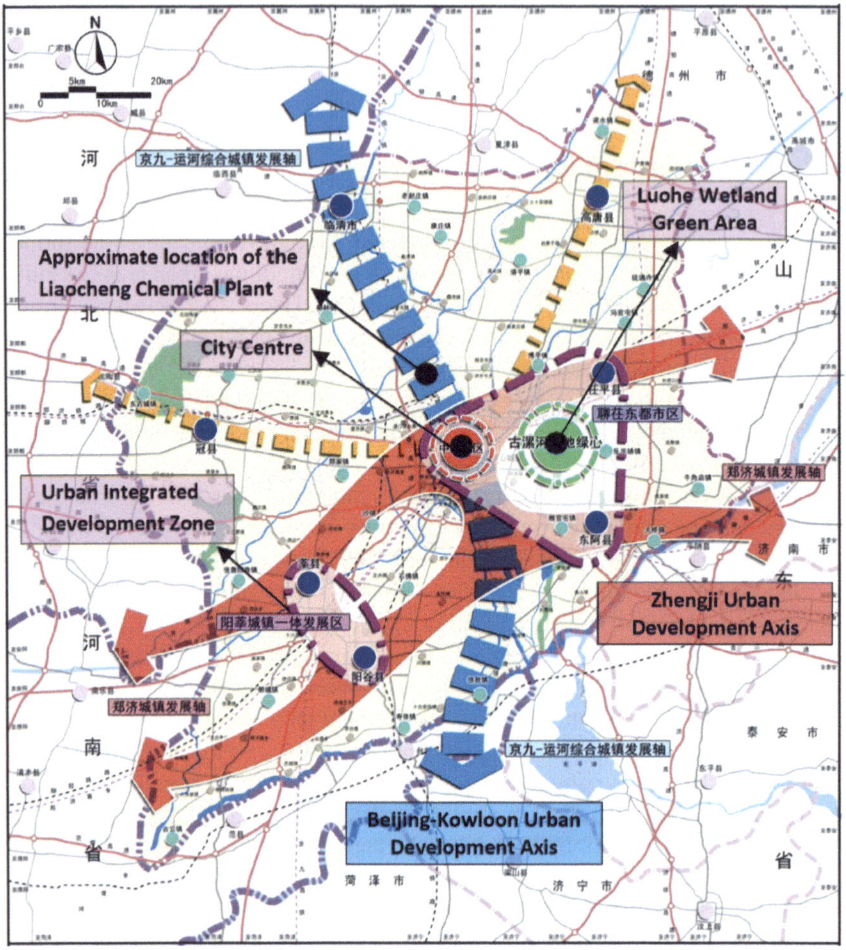

Fig. 8.3 Liaocheng City Master Plan (2013–2030); draft zoning (*Source* authors' translation and elaboration on [19])

8.5 Discussion: Brownfields in Chinese Policy

Brownfield is a new agenda into both urban land-use and environmental legal systems in China. At present, the Chinese legislative model for brownfield remediation and redevelopment is still in its infancy. This seems to be more complex when we come across the fact that there is currently no legal and nation-wide definition of what constitutes brownfields in China. However, over the past decade or so, there has been a large body of literature [1, 5, 20, 22, 34, 38] formulating a basic definition of brownfields and arguing different aspects of land remediation activities across the country. Given the lack of awareness of brownfields, these works of literature have taken critical approaches to provide a general overview of the existing situation of contaminated land redevelopment in China. The existing literature has placed a great emphasis on the vital necessity of effective and efficient management of land contamination at all levels of Chinese governments.

One of the most comprehensive surveys of the Chinese legal and regulatory framework on soil pollution prevention and control was published by the World Bank in 2010. According to this survey [39], the national/provincial brownfield legislative system in China lacks an operational management approach that could comprehensively address issues regarding remediation and redevelopment of contaminated lands. As the survey report concludes [39]:

> Currently [as of 2010], brownfield management in China is guided by an ad-hoc set of documents, rules and guidance issued by the government over the past several years, issued by MEP (formerly SEPA), in 2004 and 2008 respectively... These relevant legal provisions, however, are not systematic or consistent, and none of them are focused on land contamination. They also are too general and lack operational details and accountability deterrence provisions. Control and prevention requirements, and measures pertaining specifically to land pollution, are largely missing. It is therefore necessary to develop a specific law for prevention and control of land contamination.

However, in 2014, in response to the growing concerns over the lack of a brownfield-related management system at the national level, the Ministry of Land and Resources (MLR) in conjunction with the Ministry of Environmental Protection (MEP) released a nation-wide bulletin survey of soil pollution. This is the first and, to date, the most recent national soil pollution report in China. According to this joint survey, the overall state of the soil condition and remediation action in China is not optimistic. The report shows that over 16% of total land in China is believed to be slightly or severely contaminated [27]. Although the joint survey report of 2014 is presently the most recognized report available to the public, it does not accurately describe the actual situation of brownfields in China. First, there are doubts as to how effectively the survey has been carried out, because the sample points were evenly distributed across the entire land of China and, thus, many contaminated lands might have been missed in this survey. Secondly, many questions have been raised about the accuracy of the analytical method and tools being used by the survey. Hence, the total area of contaminated land in China is expected to be several times higher than the estimate from the national survey of 2014.

The joint survey report of 2014 also provided guidelines for local governments to take a series of measures for the prevention, control and treatment of soil pollution, following five central objectives, including [27]:

- Preparation of soil pollution prevention and control action plan
- Speeding up the legislative process of soil environmental protection
- Carrying out detailed investigation of soil pollution situation
- Implementation of soil remediation works
- Strengthening the soil environmental regulation.

Following the national bulletin survey and in response to major concerns raised by two ministries, the MEP introduced a national action plan as ‘Soil Pollution

Prevention and Control Action Plan', commonly known as 'Soil 10'. This action plan is the first, and the latest, national legislation passed regarding brownfields in China. However, in practice, Soil 10 seems to be still a very general and immature action plan, being less specific into different environmental and economic aspects of regeneration process, including both monetary issues (e.g. liability regulation and public funding), and non-monetary issues (e.g. site identification, investigation and technical assistance). Some of the major challenges and problems which have not been explicitly and comprehensively addressed by the Chinese legal and regulatory framework are discussed in the following sub-sections.

8.5.1 Identification and Characteristics of Brownfields

One of the major challenges the policy system is confronted with is the lack of unified and universally applicable methods for identification of contaminated, suspected contaminated or non-contaminated lands across China [38]. At present, there are no quantitative standards or statistical database that could help quantify, characterize and classify brownfields by providing helpful information, such as the location and size of the site, value of the land, nature of former use, period of idleness, type of contamination as well as the level of associated risk or hazardous rank. At the moment, apart from the national joint bulletin survey [27] that offers a very general overview of contamination at the national level, there is no systematic approach envisaged within the Chinese regulatory framework that supports identification of contaminated lands. This has led to a situation where the nature of several contaminated sites is still unknown and, therefore, the redevelopment of such sites may take place with no or ineffective environmental treatment. Moreover, for those sites that are known to be contaminated, the processes of site investigation, assessment and remediation often take a long time due to the uncertainty about the type and degree of pollution [25]. This situation has become widespread throughout China. The remediation project of the former site of the Liaocheng Chemical Plant is a notable example in this regard.

8.5.2 Environmental Liability Issues

Environmental and legal liabilities associated with land contamination and remediation have been always inevitable concerns for both landowners and developers. In respect of liability issues, China's regulatory system is presently in its infancy [25]. At present, there is no legislative framework at the national level in China that draws up clear and concise guidelines for identification of liability, stakeholder responsibilities and associated penalties [6, 34]. The legislation that exists at the national level regarding liability issues is pretty much fragmented into a series of laws or regulations, such as Environmental Protection Law, Land Management Law, Solid Waste

Pollution Prevention Law, Law of Water and Soil Conservation, and the Urban Real Estate Development and Management Regulations [39]. These laws or regulations are not specifically focused on liability for contaminated land management.

In the case of Liaocheng Chemical Plant site redevelopment project, the remediation expenditures were borne by the new landowner (the Liaocheng Public Security Bureau), while the polluter was the former industry. It might be that the Liaocheng Public Security Bureau paid for the remediation expenditures as part of land transfer process (given the high level of barriers to data accessibility, this cannot be certainly affirmed). There have been several brownfield remediation cases in China, in which the liability costs were deducted from the land transfer expenditures, so that the developer or new landowner bears the whole or a large amount of clean-up costs. In such cases, when it is not feasible to hold responsible parties accountable for past liabilities (such as bankrupt, closed, or restructured enterprises), the government typically intervenes by engaging in negotiations with potential developers to recover the costs of remediation through agreements involving the transfer of land. The redevelopment project of the Hongshi Paint Plant site in Beijing is one of these cases.

Despite the existing loopholes related to soil pollution liability within the national law, there are a small number of local governments in China that had already addressed liability concerns before the national government came up with Soil 10. In many states in China, the municipal government takes the overall control of environmental liability issues of contaminated sites. The most notable example in this regard is the Chongqing Municipal government. In 2007, Chongqing Municipality issued 'Chongqing Environmental Protection Regulations (CEPR)' which stipulated that soil remediation measures must be implemented by polluting industries before they may relocate or transfer the land [9]. This municipal-level regulation addressed the liability of contaminated land based on three key principles, including the 'Polluter Pays Principle', the 'Investor Benefits Principle', and the 'Land Owner Takes Responsibility Principle' [39]. However, despite considerable efforts made by the Chongqing government to resolve liability concerns, there seems to be still an ambiguity in CEPR which needs to be clarified. Contaminated sites slated for redevelopment must undergo cleanup to achieve 'pristine conditions,' with costs borne by the responsible polluters. Yet, in practice, government agencies often hesitate to enforce liability on responsible parties unless contamination poses significant risks to human health or the environment. These agencies often allow responsible parties, particularly state-owned industrial enterprises using the land, to voluntarily cleanup sites without rigorous risk assessment, oversight, or post-evaluation.

8.5.3 Funding Sources

At the national level of Chinese government, there is to date no direct funding or financing mechanisms (e.g. tax credits, low-interest loans and grants) tackling brownfield remediation and redevelopment issues. The only national-led funding sources which indirectly support the clean-up of selected types of contaminated sites in China are associated with the development of innovative treatment technologies and equipment for environmental risk assessment and remediation [25]. However, despite the lack of funding designation at the national level, there have been a small number of state and local governments in China offering subsidized funds for contaminated site remediation. This has been substantially progressing over the past years in several provincial and municipal governments across China. For example, in 2007, the Chongqing Municipal government spent 200 million RMB on 15 contaminated sites, while by 2009 the figures had risen to 800 million RMB spent on 45 sites [39].

8.5.4 Technical Issues

Since the mid-1990s, the Chinese national government has been proactive in providing several technical guidelines and standards for the prevention, control and remediation of land contamination. These national standards have provided significant technical assistance for many environmental remediation engineering and practices associated with brownfield sites. However, it seems that these standards are very fragmented and need to be reframed within the newly developed regulatory system for soil pollution prevention and control in China. In many instances, the established guidelines are not adhered to by environmental agencies and institutes [25]. Meanwhile, despite the existence of local practices and provisional standards and guidelines issued at the national level, government agencies and research institutes in China often employ or reference various methods and standards from other countries for contaminated soil risk assessment [39]. Needless to say many of these international standards diverge from the prevailing institutional structures in China, thereby yielding inconsistent and non-comparable assessment results and conclusions.

The former site of the Liaocheng Chemical Plant serves as an excellent example of China adopting an innovative approach to soil pollution control and cleanup systems. In this project, detailed assessment and remediation techniques have been developed in accordance with several national standards and technical guidelines, such as the 'Groundwater Quality Standard', 'Site Soil Environmental Risk Assessment Screening' and 'Technical Guidelines for Risk Assessment of Contaminated Sites'. After a comprehensive process of soil and groundwater analysis, both the soil and groundwater system of the Liaocheng Chemical Plant were found to be highly contaminated. In the final proposal made by the EPPI [10], the report came up with some technical solutions for soil contamination control and treatment. Some of proposed remediation techniques are innovative and deemed as emerging technologies in the marketplace of soil clean-up, such as the 'High Temperature Sintering Technology' and 'Phytoremediation Technology'. However, it must be noted that despite the high degree of groundwater pollution found at the former site of Liaocheng Chemical Plant, no technical solution has been yet offered for the control and clean-up of contaminated groundwater system. Unlike soil and surface water pollutions, groundwater pollution is more challenging to detect and control, often leading to prolonged persistence of the problem.

8.6 Conclusion

Similar to many developed and developing nations, brownfield land in China can be viewed as a legacy of two marked and inter-related trends, urban expansion and regional relocation of industries. These trends are still underway in many Chinese cities. Many cities are rapidly growing, and several post-industrial (and often contaminated) sites have now become available for redevelopment owing to the closure and relocation of industries. This is the scenario of how the former site of the Liaocheng Chemical Plant has emerged as a brownfield site. Similar to the case of Liaocheng, there are several former lands in many traditional industrial-base cities, e.g. Beijing and Shanghai, that have received legal and technical recognition for land decontamination. Despite individual attempts made by the lower tiers of local government, real estate developers, investors or individual landowners, there is still no widespread recognition for a strategic and effective approach to encourage the regeneration of contaminated sites in China. This has led to many failed remediation practices over the past few years posing a serious threat to human health and environment in China, particularly in densely populated areas. The most notable example is the soil contamination incident of the Changzhou Foreign Languages School in Jiangsu Province in 2015 [21]. The school's new site was contaminated by chemical factories that spread it further afield. The contaminated land was not treated prior to reuse and, as a result, many students became sick due to the existence of toxic chemicals in the soil and groundwater system.

Land contamination and remediation in China seems to be a common concern for both land use planning and environmental protection systems. Over the past few years, positive signs of moving forward in respect of land contamination issues appeared in Chinese legal and regulatory system. Nevertheless, it seems that there is still a lack of consensus on how to tackle various issues associated with remediation and redevelopment of contaminated lands, such as the identification of sites, legal liability, technical issues and financing mechanisms. Despite the national government's attempt to encourage the regeneration of contaminated land, there are still several elements of ambiguity and inadequacy in the national act, i.e. Soil 10. It is important to note that, in comparison with the upper level of governments, the local and provincial governments have a closer and more effective control of remediation practices in China. However, several key issues, such as funding allocations and provision of technical standards and guidelines, are still sorted out at the national level. This disconnection between multiple levels of government has presented significant challenges to brownfield remediation and reuse practices. The response to these challenges must be viewed at every level of government, and requires the national government sets appropriate brownfield-related legislative targets, and accordingly creates a detailed operational framework. Given the rapid pace of development and growth in China, it is imperative that the Chinese government accelerates the design and implementation processes to achieve these legislative targets.

Acknowledgements The authors are grateful to the informants for their time in discussing the issue of brownfield regeneration in China. Special thanks are also extended to Prof Xueliang Yuan and his research team at Shandong University for their assistance in identifying key informants.

Declaration Statement This chapter draws substantially from two chapters of a PhD dissertation authored by the primary author, under the supervision of the co-author.

References

1. Cao K, Guan H (2007) Brownfield redevelopment toward sustainable urban land use in China. Chin Geogr Sci 17:127–134
2. Champion T (2001) Urbanization, suburbanization, counter urbanization and reurbanization. Handbook Urban Stud 160(1):143–161
3. Chang G, Jiang C, Chang K, Alam B (2013) Land prices and intracountry industrial relocation in China. Chin Econ 46(2):54–73
4. Chen X (2014) Steering, speeding, scaling: China's model of urban growth and its implications for cities of the global south. In: The Routledge handbook on cities of the global south. Routledge, pp 155–172
5. Cheng F, Geertman S, Kuffer M, Zhan Q (2011) An integrative methodology to improve brownfield redevelopment planning in Chinese cities: a case study of Futian, Shenzhen. Comput Environ Urban Syst 35(5):388–398
6. Chiang SYD, Gu Q (2015) Brownfield sites remediation technology overview, trends, and opportunities in China. Remediat J 25(3):85–99
7. Démurger S, Sachs JD, Woo WT, Bao S, Chang G, Mellinger A (2002) Geography, economic policy, and regional development in China. Asian Econ Papers 1(1):146–197
8. Dongchangfu (2024) Overview of Dongchangfu district (in Chinese). http://www.dongchangfu.gov.cn. Accessed 15 July 2024
9. EFB (2024) Chongqing environmental protection regulations. Chongqing Environmental Protection Bureau http://www.cepb.gov.cn/. Accesssed 01 July 2024
10. EPPI (2017) Liaocheng sulfuric acid development plan—detention centre, office buildings and armed police land use (in Chinese). Environmental Protection and Planning Institute
11. Fang Y, Pal A (2016) Drivers of urban sprawl in urbanizing China—a political ecology analysis. Environ Urban 28(2):599–616
12. Han W, Zhang X, Zheng X (2020) Land use regulation and urban land value: evidence from China. Land Use Policy 92:104432
13. He C, Zhou Y, Huang Z (2016) Fiscal decentralization, political centralization, and land urbanization in China. Urban Geogr 37(3):436–457
14. Jacek G, Rozan A, Desrousseaux M, Combroux I (2022) Brownfields over the years: from definition to sustainable reuse. Environ Rev 30(1):50–60
15. Jiang X, Lu X, Liu Q, Chang C, Qu L (2021) The effects of land transfer marketization on the urban land use efficiency: an empirical study based on 285 cities in China. Ecol Ind 132:108296
16. Kano L, Tsang EW, Yeung HW-C (2020) Global value chains: a review of the multi-disciplinary literature. J Int Bus Stud 51(4):577–622
17. Kuang W, Liu J, Dong J, Chi W, Zhang C (2016) The rapid and massive urban and industrial land expansions in China between 1990 and 2010: a CLUD-based analysis of their trajectories, patterns, and drivers. Landsc Urban Plan 145:21–33
18. LCG (2024) Liaocheng city profile (in Chinese). http://www.liaocheng.gov.cn. Accessed 05 Jul 2024

19. LCPB (2024) Liaocheng city master plan (2013–2030). Liaocheng City Planning Bureau. www.lcghj.gov.cn/. Accessed 12 Jul 2024
20. Li X (2011) Brownfields in China: how cities recycle industrial land. Massachusetts Institute of Technology
21. Liu C, Song W, Zhou C (2017) Unsuccessful urban governance of brownfield land redevelopment: A lesson from the toxic soil event in Changzhou, China. Sustainability 9(5):824
22. Liu Y, van Oort F, Geertman S, Lin Y (2014) Institutional determinants of brownfield formation in Chinese cities and urban villages. Habitat Int 44:72–78
23. Liu Z, Liu S, Qi W, Jin H (2018) Urban sprawl among Chinese cities of different population sizes. Habitat Int 79:89–98
24. Longo A, Campbell D (2017) The determinants of brownfields redevelopment in England. Environ Resour Econ 67:261–283
25. Mehdipour A (2020) Understanding brownfield regeneration in Iran through the lens of international experience
26. Mehdipour A, Kellett J, Palazzo E, Larbi M (2024) Policy integration for brownfield regeneration: an analytical tool. Sustain Horiz 10:100085
27. MEP (2014) Investigation on the situation of soil pollution in China (in Chinese). The National bulletin by the Ministry of Environmental Protection & the Ministry of Land and Resources
28. MEP (2024) Notice on strengthening pollution prevention and control of industrial enterprises in restraining, relocating, redevelopment and utilization of site (in Chinese). Ministry of Environmental Protection. http://www.mep.gov.cn/. Accessed 10 Jul 2024
29. MEP (2024) Notice on the work of preventing and treating environmental pollution in the process of enterprise relocation (in Chinese). Ministry of Environmental Protection. http://www.mep.gov.cn/. Accessed 10 Jul 2024
30. Oliver L, Ferber U, Grimski D, Millar K, Nathanail P (2005) The scale and nature of European brownfields. Paper presented at the CABERNET 2005-international conference on managing urban land LQM ltd, Nottingham, UK, Belfast, Northern Ireland, UK
31. Pacione M (2009) Urban geography: a global perspective: Routledge
32. Reba M, Seto KC (2020) A systematic review and assessment of algorithms to detect, characterize, and monitor urban land change. Remote Sens Environ 242:111739
33. Schneider A, Mertes CM (2014) Expansion and growth in Chinese cities, 1978–2010. Environ Res Lett 9(2):024008
34. Sun Y, Li H, Lei S, Semple KT, Coulon F, Hu Q et al (2022) Redevelopment of urban brownfield sites in China: Motivation, history, policies and improved management. Eco Environ Health 1(2):63–72
35. USEPA (1995) Brownfields national partnership action agenda. United States Environmental Protection Agency, Washington, DC
36. Wei YD (2001) Decentralization, marketization, and globalization: The triple processes underlying regional development in China. Asian Geogr 20(1–2):7–23
37. Wu F, Xu J, Yeh AG-O (2006) Urban development in post-reform China: state, market, and space. Routledge
38. Xie H, Lin M, Zhou C, Zhang Y, Zhou L (2014) Identification of brownfields in China: concept, procedure and practice. J Food Agric Environ 12(3–4):349–354
39. Xie J, Li F (2010) Overview of the current situation on brownfield remediation and redevelopment in China. World Bank
40. Ye L, Wu AM (2014) Urbanization, land development, and land financing: evidence from Chinese cities. J Urban Aff 36(sup1):354–368
41. Zhang T (2000) Land market forces and government's role in sprawl: the case of China. Cities 17(2):123–135
42. Zhao P (2010) Sustainable urban expansion and transportation in a growing megacity: consequences of urban sprawl for mobility on the urban fringe of Beijing. Habitat Int 34(2):236–243
43. Zhou Y, Ma LJ (2000) Economic restructuring and suburbanization in China. Urban Geogr 21(3):205–236

Chapter 9
Fintech as Developmental Catalyst or Exploitative Mechanism? A Case Study of Moove in Lagos, Nigeria

Chen Wu and Ding Fei

Abstract Situating in the scholarly debate on whether Fintech serves as a development catalyst for African leapfrogging or a new mechanism for capitalist exploitation, the chapter asks how Fintech—as a socio-technical infrastructure—reshapes urban economy and experiences of urban living. The empirical case focuses on the Nigerian ride-hailing sector, examining the collaboration between Fintech companies and ride-hailing platforms. Using a case study of Moove, a Fintech company, and Uber, a global giant of ride-hailing platform, we explore how such partnerships transform the urban mobility landscape in Lagos on the one hand, and tremendously influence drivers' livelihoods and wellbeing, on the other. Specifically, we argue that Fintech expands vehicle financing solutions, enabling more Nigerians to participate in the digital economy. However, the collaboration between Fintech companies and ride-hailing platforms also exacerbates employment vulnerability in times of economic hardships and policy change in Nigeria. The case sheds light on the complexities and socio-economic implications of Fintech-platform collaborations in Nigeria's ride-hailing sector and inform future scholarly conversation about and policy intervention towards urban mobility and financial inclusion.

Keywords Platform · Mobility · Fintech · Urbanism · Nigeria · Labor relationship

9.1 Introduction

The emerging wave of digital transformation across Africa is bringing new dynamics to human interaction and urban life in African cities, reshaping the way that the city functions and is experienced. Since Uber first launched in Johannesburg, in 2013, an increasing number of global ride-hailing platforms and local players have joined the African market. A recent report projects that Africa's shared mobility market will double in size by 2030, reaching $8 billion [3]. This growth is anticipated to

---

C. Wu (✉) · D. Fei
Department of City and Regional Planning, Cornell University, Ithaca, NY, USA
e-mail: cw948@cornell.edu

G. Siew et al. (eds.), *Diversity as Catalyst: Economic Growth and Urban Resilience in Global Cityscapes*, Urban Sustainability, https://doi.org/10.1007/978-981-97-8776-0_9

create an additional 550,000 income-earning opportunities on the continent. The expansion of the ridesharing platforms introduces not only a nascent alternative for transportation, but new narratives and imaginaries of social-spatial connection and urban entrepreneurialism across the continent [5, 24]. These platform-based businesses often operate differently across political economic contexts, influenced by a country's legislation on tech corporations, governance regime for foreign entities, as well as existing urban mobility landscapes.

The ride-hailing sector in Nigeria marks a vibrant and rapidly evolving terrain. Since the introduction of major players like Uber in 2014 and Bolt (formerly Taxify) in 2016, the sector has experienced substantial growth, both in terms of service offerings and market penetration. By 2024, the annual revenue is expected to reach $252.7 million and maintain double-digit growth over the next decade [27]. Several factors have driven this growth. First, with the largest population in Africa, Nigeria's youthful population and high unemployment rates create a robust supply and demand for gig-based employment. Second, increasing smartphone and internet penetration have provided the necessary infrastructure for digital transformation. Third, Nigeria's liberal policy in currency exchange and the legibility of market entry for foreign investors makes its digitalization market be more attractive to global venture capitals (VCs) [21].

After a decade of growth, Nigeria's ride-hailing sector has experienced new opportunities and challenges [11]. The sector has become increasingly competitive, with traditional giants such as Uber and Bolt facing increasing competition from newer entrants like Indriver and Rida. The sector is also witnessing a shift towards more sustainable practices with the introduction of electric vehicles (EVs). This transition is supported by the rising demand for cleaner and more cost-effective transportation solutions. EV ride-hailing platforms, such as EV Taxi, are at the forefront of this shift, aiming to reduce carbon emissions and operational costs [3]. Moreover, ride-hailing platforms are increasingly seen as more than just transportation service providers, but are critical infrastructures enabling financial inclusion and enhancing urban mobility enhancement [7]. This implies a complex network of stakeholders including rides-hailing platforms, online payment platforms, users, drivers, the government, vehicle manufacturers, local authorities, Fintech companies, and banks.

The chapter centers on one of the crucial and emerging players in the ride-hailing sector—Fintech companies. It charts out the collaboration between Fintech companies and ride-hailing platforms in sustaining the ride-hailing sector. This nexus is often seen as a necessary strategy of localization to enlarge the number of drivers in a country with low vehicle ownership. In Nigeria, whole foreign-holdings companies without local partnerships are prohibited from owning physical assets such as vehicles and land. This regulation incentivizes ride-hailing platforms to seek collaboration with local actors who can offer vehicles and finance to expand the potential pool of drivers. Yet, how does the collaboration materialize? What are the impacts on workers and their livelihoods? To answer these questions, the chapter adopts a case study approach to unpack the collaboration between Moove, a rising Fintech company, and Uber, the largest ride-hailing platform in Nigeria. Our analysis transcends sectorial or firm-level perspectives to illuminate the adverse impacts of ride-hailing on workers and their livelihoods. In doing so, it aims to provide a critical urban interpretation of the consequences and stakes associated with the intertwined growth of the Fintech and ride-hailing economies. The actual lived experiences of mobility, work, and life highlight the long-term socio-political ramifications of persistent urban challenges. They also reveal the efforts and hurdles faced by platforms, Fintech companies, and drivers' unions as they strive to forge balanced collaborations to address emerging challenges in the sector.

9.2 Fintech and Ride-Hailing in Africa

The impact of Fintech in Africa has been a subject of intense debate over the last decade. Advocates celebrate Fintech as a catalyst for economic transformation in Africa, driven by rising foreign direct investments. They argue that it enables African nations to leapfrog traditional infrastructural constraints, addressing the continent's weak banking systems and limited financial capacity [26]. This rise is propelled by increasing mobile adoption, a large unbanked population, and a relatively young demographic. However, critics warn that this optimistic portrayal of Fintech masks underlying exploitation and social disparities. Research has indicated that digital platforms function as performative infrastructures, compelling informal populations to adapt to predetermined commercial models [19]. Others have pointed to the unequal deployment of Fintech across different social strata and between urban versus rural settings [2, 4]. These scholars advocate for a thorough examination of the complex interplay between Fintech, urban economy, and labor exploitation that characterizes Africa's "digital turn" [17, 18].

This chapter delves into this scholarly debate on whether Fintech acts as a catalyst for Africa's developmental leapfrogging or as a tool for capitalist exploitation. It examines the role of Fintech as a socio-technical infrastructure in transforming the urban economy and the everyday urban living. By analyzing the concurrent expansion of the Fintech and ride-hailing sectors in Africa and Nigeria in particular, we investigate how these emerging partnerships influence and are influenced by the contextually specific and uneven conditions of growth and marginalization; and how these dynamics introduce new opportunities and challenges for urban dwellers.

In 2020, the Fintech sector in Africa drew the largest share of VC investments, with 28% of all VC deals across the continent focusing on Fintech companies [12]. South Africa, Uganda, Nigeria, and Kenya are the primary hubs for Fintech headquarters [26]. The emerging Fintech sector greatly facilitates the digital transformation in all kinds of industries, like banking, transportation, and manufacture. As they are unlocking new business models beyond financial services, they also exact significant influence on the activity pattern of urban daily life [16]. Three impacts are commonly outlined in broader narratives about Fintech: stimulating economic activities, creating multiplier effects, and driving progress towards development goals. Yet, while often being portrayed as a catalyst for economic growth, Fintech companies rarely operate in singular manners. Their catalyzing effects often manifest through the collaboration with other entities such as traditional banks and ride-hailing platforms.

Fintech companies and ride-hailing platforms are interconnected in several aspects to shape everyday mobility and livelihoods in African cities (Fig. 9.1). First, Fintech provides payment solutions for riders that are crucial for ride-hailing platforms in Africa, where cashless transactions are increasingly preferred for reasons of security and convenience. Flutterwave, based in Kenya but operating across several African countries, offers a payment gateway that ride-hailing services can integrate into their apps to facilitate seamless and secure payment transactions. In Nigeria, Flutterwave has partnered with Uber to enable a feature called "Uber Cash" which allows riders to prepay for their upcoming trips.

Second, across Africa, many drivers on ride-hailing platforms might not have access to traditional banking services. Fintech companies step in to fill this gap by offering financial products such as microloans, insurance, and savings plans tailored specifically to drivers. Increasingly, Fintech firms often collaborate with ride-hailing platforms to offer vehicle loans and leases under more accessible conditions than those available through traditional banks. In addition to the Moove-Uber case discussed in this chapter, another example involves SafeBoda and M-KOPA in Uganda. SafeBoda, a motorcycle ride-hailing service, partnered with M-KOPA, a Fintech company known for providing "pay-as-you-go" financing solutions. This partnership allows SafeBoda drivers to access motorcycles on a lease-to-own basis, facilitated by M-KOPA's financing model. Drivers make daily or weekly payments towards owning the motorcycle, with the payment amounts tailored to their earning patterns.

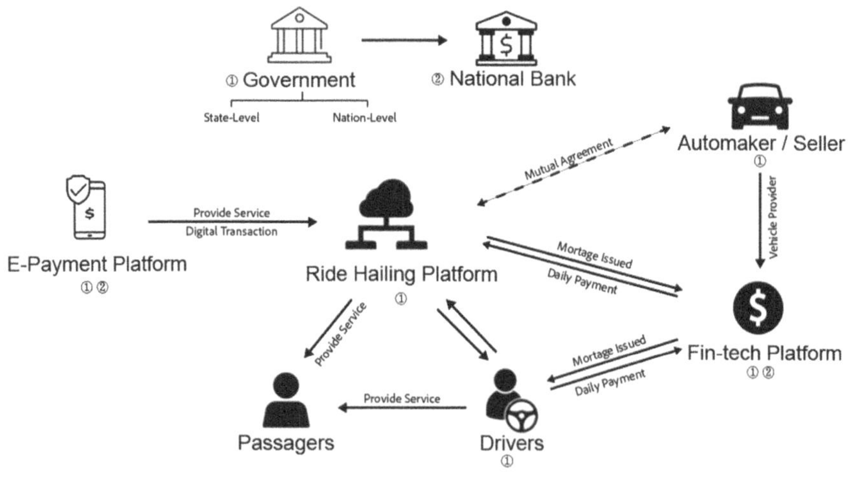

Fig. 9.1 Simplified web of relationship of Nigeria ride-hailing sector

Third, Fintech companies leverage the vast amount of data generated by ride-hailing platforms to offer customized financial services. By analyzing patterns in earnings and spending, Fintech can provide personalized financial advice, better risk assessment for loans, and tailored insurance packages, enhancing the financial well-being of drivers and operators. One notable example is the collaboration between Branch International and Uber in Kenya. Branch, a Fintech firm, uses data science to analyze ride-hailing drivers' earning patterns and other data points collected via the Uber app. Another example involves Paystack, a Fintech companies that collaborated with Bolt in Nigeria. By processing payment for Bolt, Paystack also provides Bolt with analytics, such as trends in customer payment behaviors, success rates of different payment methods, and geographical payment patterns. Such detailed analytics help Bolt make informed decisions about where to focus their marketing efforts, optimize their operations, and enhance the overall user experience by understanding better how customers prefer to pay for their services.

In addition to these various aspects of collaborations, Fintech firms and ride-hailing platforms also navigate complex regulatory regime by adapting to the specific legal and operational requirements of each country. They often engage in active dialogue with different levels of governments, participate in shaping policy through advocacy, and ensure compliance through strategic planning and technological adaptation. SafeBoda, a motorcycle ride-hailing service operating in Uganda, provides a good example of how to navigate regulatory regimes. Faced with stringent regulations aimed at formalizing the informal boda-boda (motorbike taxi) industry, SafeBoda worked closely with the Ugandan government. This collaboration helped shape the development of safety regulations that were feasible for riders and beneficial for public safety. SafeBoda introduced safety features such as helmets for both riders and passengers, reflective jackets, and a rider training academy to ensure compliance with these new regulations. This proactive approach not only helped SafeBoda align with government expectations but also positioned the company as a leader in promoting safety and professionalism in the industry.

The interactions among ridesharing platforms, Fintech companies, and drivers are often portrayed as a collaborative network. Ridesharing platforms connect drivers, who are registered but not employed by the platform, with Fintech companies. These Fintech companies function as lenders, offering vehicle loans to the drivers. Drivers use the ridesharing platform to meet passenger demand and, with the earnings, repay their vehicle loans to the Fintech companies. As the chapter will show, while Fintech companies claim to help drivers acquire their own vehicles, this arrangement has inadvertently led to increased tensions and conflicts between the drivers, the Fintech companies, and the ridesharing platforms.

9.3 Case Study: Uber-Moove Partnership in Nigeria

Uber launched its business in Lagos and Abuja in 2014, as its commencement to the Nigeria market. Initially starting in Johannesburg, South Africa in August 2013, Uber expanded rapidly across Africa continent and has reshaped the overall ride-hailing market landscape to be more digitalized and platform-based. Confronting competition and local protests, Uber integrated the strategy of localization into its business model to better accommodate itself to the local contexts. For example, Uber introduced cash payments in Nairobi, as an alternative when the online transaction fails to work. To address the challenge of low vehicle ownership in Africa, Uber actively collaborates with local banks and mortgage agencies to facilitate prospective drivers to own personal vehicle to start their business. Moove is the monopoly Uber partner in Nigeria to provide vehicle loan and mortgage services in Nigeria. By charting out the logics and modalities of such collaboration, this section explores ways that Fintech and ride-hailing capital influence the landscape of urban mobility and livelihoods.

9.3.1 Financial Access

Mooves emerges to provide innovative vehicle financing solutions, primarily targeting "mobility entrepreneurs" such as Uber drivers. Comparing to conventional financing institutions, Moove offers a more friendly loan approval criteria, especially for individual lender. Most banks and lenders are likely to reject the vehicle mortgage applications from prospective drivers due to the perceived high risk that freelancers may be unwilling or unable to repay their debt. Moove addressed this challenge by implementing a revenue-based financing model, where Moove is arranged to receive payments first. Under this model, Moove negotiates agreements with ride-hailing platform, allowing the company to monitor each vehicle's performance, track the number of trips made, and measure the revenue generated. The shared data, greatly enhancing the information transparency among the marketplace partners, allows Moove to predict and prevent the possible default.

Although the process of reviewing and verifying the eligibility of drivers applying for Moove mortgages remains opaque, it is acknowledged by both the platform and drivers that Moove has improved access to revenue-based vehicle financing. By easing the barriers for drivers to get financed, Moove significantly lowers the financial threshold for individual entrepreneurs to enter ride-hailing sector as a service provider. Meanwhile, the ride-hailing platforms also rely on Moove to support the growth of market in emerging economies. The rise of Moove is deemed to release the potential of prospective drivers and encourage them to work in full-time. The growth in the numbers of operating drivers would in turn accelerate further market penetration. Besides, owing to the restrictions in regulations and risk of compliance, the ride-sharing platforms prefer to collaborate with private financing companies, rather than commercial banks. Currently, as Uber's largest vehicle financing partner in the world, Moove has built a profound partnership with Uber while their collaboration could be reached out globally in Europe and Asian.

9.3.2 Safe and Clean Vehicles

The rise of Moove, in collaboration with Uber, aims to revolutionize the safety and reliability of ride-hailing services by improving vehicle conditions in Nigeria—a country plagued by high road fatality rates and a market saturated with unsafe, second-hand vehicles. Recognizing an opportunity to enhance driving safety and vehicle quality, Moove provides brand-new vehicles to drivers, thereby circumventing the risks associated with the prevalent influx of compromised second-hand cars. These are often salvaged vehicles that have failed safety and emissions tests abroad, hastily repaired and sold cheaply due to lower import duties compared to fully assembled cars.

Uber enforces stringent requirements for vehicles, including maximum operating age, comprehensive insurance coverage, specific vehicle types, and mandatory pre-service inspections. Despite these regulations, the system is sometimes undermined by fraudulent documentation intended to bypass safety checks. Moove's partnership with Uber fortifies the scrutiny of vehicle conditions, making it more practical and reliable. Moove not only sells new vehicles directly to drivers but also conducts thorough due diligence and inspections after purchasing vehicles from manufacturers. This ensures that both the drivers and the platform receive vehicles in optimal condition.

In addition, Moove is authorized to monitor the vehicles' performance and location, allowing regular checks to ensure that vehicles are well-maintained and operated efficiently. This ongoing monitoring helps to guarantee consistent revenue generation and aligns with Uber's goal of minimizing risks associated with vehicle conditions, thus sustaining safe and reliable ride-hailing services.

9.3.3 Car Ownership

In many African countries, low vehicle ownership rates pose significant challenges for the growth of ride-hailing platforms. The 2021 Nigeria Multiple Indicator Cluster Survey (MICS) revealed that only 10.4% of Nigerian households have at least one member who owns or has access to a vehicle. This contrasts sharply with the 91.7% vehicle ownership rate among U.S. households, highlighting a substantial barrier to building a stable and sufficient supply of ride services in emerging markets.

The success of ride-hailing, as a new business model on the African continent, hinges on the ability to offer sufficient, convenient, and affordable transportation options, to foster habitual use, and ultimately expand the market. To address the vehicle ownership challenge, ride-hailing platforms have strategically partnered with local lending agencies. These collaborations enable potential drivers to acquire vehicles and start their own ride-hailing businesses. This localization strategy has proven effective. For example, after Moove launched its services in 2020, there was a notable surge in driver applications, indicating that the approach met Uber's expectations and successfully tapped into the market's potential.

9.3.4 Investor Relations

As an emerging Fintech start-up addressing challenges for both ride-hailing platform and drivers, Moove have garnered attention from the global VC investors. The confidence and enthusiasm from these investors have stimulated dramatic growth over the past four years. This growth momentum persists despite the challenges posed by the pandemic and economic downturns. By March 2024, Moove had completed two rounds of VC financing from investors in UK, USA, UAE and European. The fundings, incorporating debt and equity have accumulated to about $380 million.

In March 2024, following its Series B funding round of $100 million, Moove's post-money valuation reached $750 million [15]. This round, which included significant contributions from Uber and the sovereign wealth fund Mubadala (the latter having led Moove's previous Series A2 round in 2023), was primarily aimed at expanding into new African markets and advancing the electrification of its fleet.

Two important sources of investment marked Moove's success in the path of financing. In 2021, Moove raised $23 million in Series A funding. This round was led by Speedinvest and Left Lane Capital, with participation from DCM, Clocktower Technology Ventures, and 12 other investors, bringing Moove's total funding to $68.2 million. This initial capital injection, following its startup launch in June 2020, highlighted broad interest from global VC firms. In 2022, Moove secured an additional $105 million in Series A2 financing to support its expansion into seven new emerging markets in Europe, Asia, and the Middle East, with continued backing from Speedinvest, Left Lane Capital, and other ventures. These investments underscored Moove's potential to bridge the vehicle financing gap for prospective drivers globally.

Since its inception, Moove has not only expanded its vehicle offerings to include electric vehicles and emphasize empowering female entrepreneurs but also broadened its geographical footprint. Moove has expanded its operations across Africa and into other regions, including a notable expansion into Nairobi, Kenya. This move into the Kenyan market marked Moove's entrance into East Africa and involved partnerships with local logistics companies like Sendy, broadening their vehicle financing to include motorcycles and trucks for various delivery services.

9.4 Economic and Policy Changes in Nigeria

Since the pandemic, Nigeria's economy and policy regime has faced multiple challenges, which has not only influenced digital sector players such as Uber and Moove, but also individual drivers employed in the country. This section reviews the important changes economic and policy space and discusses their impacts on the market conditions and people's livelihoods.

9.4.1 Economy Challenge: Inflation and Foreign Exchange Rate

At the end of 2023, Nigeria's inflation rate, measured by the average annual positive variation in Customer Price Index, reached to about 25%. It eroded consumer purchasing power and increased the cost of living, affecting both businesses and households. In 2024, the national average price of rice soared, doubling from around ₦700 (approximately $0.52) per kg in 2023 to ₦1600 (approximately $1.20) per kg. Similarly, the price of beans tripled from about ₦600 (approximately $0.44) per kg to ₦1900 (approximately $1.40) per kg [30]. The World Bank estimated the rising inflation and weak earnings pushed 10 million Nigerians into poverty in 2023.

The rising inflation is attributed to the weakening Nigerian currency and the inadequately recovered supply chain since the pandemic [29]. According to the official exchange rate published by the Central Bank of Nigeria (CBN), the Naira has depreciated significantly, from ₦770/$ in June 2023 to ₦1597/$ in February 2024, compared to about ₦450/$ in January 2023. This sharp decline in the Naira's purchasing power in the international market has had profound domestic impacts, particularly in sectors essential for daily needs. Both the general manufacturing industry and the agricultural sector have felt the effects of this currency depreciation, leading to increased costs for imported raw materials [20].

As a response, in 2023, CBN announced reforms to unify all foreign exchange rate windows, aiming to consolidate the various exchange rate segments to increase transparency, enhance liquidity, and stabilize the foreign exchange market. The shift allows more flexibility and market-driven pricing to attract foreign investment, yet it also leads to short-term volatility as markets adjust to the new system. The volatility can impact the cost of imported goods, including necessities like food and medicine, exacerbating the conditions of inflation and affecting daily living costs.

The Fintech and ride-hailing sectors are directly impacted by these economic conditions. Inflation leads to increased operational costs. For ride-hailing, this includes rise in the cost of fuel, vehicle maintenance, and spare parts, all of which can increase the overall cost of providing services. For Fintech, inflation can increase the cost of infrastructure, such as server maintenance and cybersecurity measures. High inflation reduces consumers' purchasing power, affecting their ability to spend on services, reducing transaction volumes and lowering revenue for businesses. Fintech startups, which often rely on investor funding for growth, may find it challenging to secure investments as inflation rises and the foreign exchange rates fluctuates. Ride-hailing companies might need to adjust their pricing strategies frequently to accommodate fluctuations in fuel prices and other operational costs impacted by exchange rate changes.

Challenges experienced at sector and firm level often trickle down to individual workers. High inflation and exchange rate fluctuations can lead to increased operational costs for ride-hailing drivers, which can, in turn, reduce consumer demand for ride-hailing services due to increased fare prices. Ride-hailing drivers would experiencing severe financial crises as their net income falls short of a livable standard due to rising living costs. Consequently, some drivers have protested, arguing that the working arrangements with the vehicle financing platform, particularly the weekly remittance requirements, exacerbate their stress rather than alleviate it [14].

9.4.2 Policy Shift: Fuel Subsidy Removal

Fuel subsidies in Nigeria were introduced in the 1970s. The government began the practice of subsidizing petrol costs to stabilize domestic prices and shield consumers from the volatility of international oil markets. This was also a period of significant oil revenue for the country, which made it financially feasible to subsidize fuel prices. The subsidies became institutionalized with the promulgation of the Price Control Act in 1977 under General Olusegun Obasanjo's regime. This law was intended to cushion the effects of inflation, which was affecting global economies due to sharp increases in energy prices at the time. It made it illegal for some products, including petrol, to be sold above regulated prices.

While initially popular, the sustainability of fuel subsidies has been questioned over the years [23]. Subsidies have consumed a significant portion of the national budget, diverting funds from other critical areas such as infrastructure and healthcare. For example, in 2022, fuel subsidies accounted for about 24% of Nigeria's budget. Moreover, they have led to corruptions and economic inefficiencies, including fuel smuggling and reduced incentives for investment in alternative energy sources. In recent years, various administrations have proposed plans to phase out these subsidies, citing the unsustainability of the subsidy regime and the need to reallocate financial resources to more productive sectors of the economy [1]. The World Bank and other international financial institutions have also recommended the removal of fuel subsidies to improve Nigeria's economic health.

Earlier attempts to remove or reduce fuel subsidies have led to widespread protests and political unrest, most notably in 2012, when the government's attempt to remove subsidies led to massive protests across the country. The removal of subsidies was seen by many as necessary for economic reform but remained highly unpopular among the general populace due to the immediate impact on fuel prices and the cost of living. The recent cancellation of fuel subsidies was announced by President Bola Ahmed Tinubu shortly after his inauguration in May 2023. It has led to a significant increase in fuel prices across the country, a spike in the pump price of petrol from about ₦780 a gallon (approximately $1) to ₦2160 a gallon ($2.80). The societal impact has been profound, especially for lower-income and rural households, as the cost of transportation and goods has risen sharply.

From a strategic perspective, the government intends to use the savings from subsidy removal to invest in infrastructure and social welfare programs that could provide broader benefits to the population. However, without careful planning to mitigate the negative impacts on the less affluent and gain public trust through transparent and effective use of the redirected funds, the policy decision has increased the plight of the ordinary citizens. The government faces challenge of balancing fiscal health with social stability. Effective communication, gradual implementation, and compensatory measures are crucial to navigate the political and social sensitivities associated with this reform.

9.4.3 Policy Initiative: Guidelines for Online Hailing Business Operation of Taxi

In August 2020, the Lagos State Government, through the Lagos State Ministry of Transportation (LSMT), issued the Guidelines for Online Hailing Business Operation of Taxi in Lagos State ("the guidelines"). The primary goals of these guidelines were to enhance traffic safety, clarify accountability among stakeholders, and establish a clear licensing procedure. It is worth noting that the guidelines focus extensively on the relationship between local municipalities and ride-hailing platforms as captured in Fig. 9.1.

However, Moove and its drivers, as critical stakeholders in the ride-hailing business, were significantly impacted by the subsequent enforcement actions. Although financial services fall outside the jurisdiction of the Ministry of Transportation, the interconnections between Uber, the government, and potentially Moove's business cannot be ignored. The guidelines require all operators and platforms to provide unrestricted access to their databases for the LSMT and its agencies. While Uber agreed to share only the "required data" with the government, LSMT escalated the situation by threatening Uber and began impounding its drivers' vehicles [13].

Given the monopoly relationship between Moove and Uber, and because Moove vehicles are easily identifiable by their specific models, Moove drivers found themselves at the center of this conflict. Enforcement officers could readily identify and impound these vehicles, making Moove drivers the most adversely affected group. They were left out of the dialogue, and the impoundments jeopardized their only source of income, pushing many towards mortgage default. In the development of regulations for the ride-hailing industry, drivers often emerge as the most vulnerable group, facing the brunt of policy changes and regulatory shifts.

In addition to data sharing, the guidelines specify the licensing fees charged to the platform and a 10% service tax on each transaction, which is paid by passengers to the operators. This tax directly affects the daily revenue that drivers earn, potentially reducing their net income as more players partake in revenue distribution. Without a corresponding increase in ride-hailing service prices, drivers ultimately bear the burden of this policy enforcement. Since the service tax is collected by the LSMT, it effectively becomes an additional local tax item, separate from the personal income tax collected by the Federal Inland Revenue Service (FIRS) [9].

9.5 Platform Workers: Impacts and Responses

A recent survey by Cieslik et al. [6] offers a profile of ride-hailing drivers in Nigeria, revealing that half of these drivers operate leased vehicles, while a mere 1% are women. This gender disparity is attributed to the unfriendly and harsh working conditions that pervade the sector. Set against a backdrop of significant economic and policy shifts, this section delves into the vulnerabilities and challenges faced by platform workers, examining the elements of precarity and the forms of resistance they employ. This analysis not only highlights the economic dependencies created by the platform economy but also uncovers the social dynamics and struggles for rights within this rapidly evolving sector.

9.5.1 Precarity and Resistance

Platform workers—ride-hailing drivers—are especially vulnerable to economic volatility and policy changes in Nigeria. They often find themselves in a state of informal employment, which comes with significant job insecurity [6]. These workers are typically classified as independent contractors rather than employees, which exempts the companies from offering job security, benefits, or standard labor protections that are otherwise provided to formal employees. This classification impacts their financial stability as they are responsible for their own operational costs, including fuel, vehicle maintenance, and insurance, without any guaranteed income. The lack of formal employment terms also means that these workers face high variability in earnings and have no protection against job termination or changes in corporate policies that might unfavorably affect their income. As such, while ride-hailing apps provide critical employment opportunities in a country with high unemployment rates, they also contribute to a precarious job landscape where workers are left without the safety nets that formal employment would provide. Meanwhile, the absence of industry-specific regulations means that ride-hailing companies have considerable latitude in setting fares and commission rates, often without sufficient input from or consideration of the drivers. This regulatory gap not only places drivers in a vulnerable economic position but also leaves them with little recourse in disputes with ride-hailing platforms.

Despite of their structural vulnerability and precarity, workers have actively engaged in union/association formation and protest/strike organization to develop alliances and make their voices heard [6] (Table 9.1). In 2020, in responding to the enforced Guidelines for Online Hailing Business Operation of Taxi, the drivers in Lagos State started an organized protest to against the proposed service tax. The protest successfully leads to an amendment of guidelines. In the amendment, the service tax got decreased from 20% to 10% and the license-renewal fee was lowered. In 2021, On April 19, Uber drivers in Lagos, organized under the Professional E-hailing Drivers and Partners Association (PEDPA), initiated a one-week strike to protest exploitative conditions. They demand fare price adjustments to accurately reflect their operational costs and a reduction of the commission from 25% to 10%.

In January 2023, ride-hailing drivers formed the Amalgamated Union of App-Based Transport Workers of Nigeria (AUATWON) to advocate for better working conditions and protections. This union was recognized officially by the Nigerian government and has been active in advocating for the rights and welfare of its members, including pushing for lower commission rates charged by ride-hailing companies and better profiling of riders to enhance driver safety.

In February 2023, thousands of Uber drivers in Lagos, registered on MOOVE went on the street to protested over continued frustration and unbearable financial burden. "We must be online for 12 h to potentially earn $28. After paying $20 to Moove and $5 to Uber, plus $4 for fuel and money for a calling card, we're often left with little to nothing", a protesting driver stated [22]. In June 2023, drivers under AUATWON planned another strike due to dissatisfaction with the responses of ride-hailing companies to their demands for a 200% price increase and a base fare increase on their platforms. This was driven by the high operational costs that the drivers faced, which were not adequately covered by the existing fare structures.

Following the fuel subsidy removal in August 2023, AUATWON launched serval protest campaign for increasing the expense of passengers and decrease the monthly debt service charged by Fintech companies. The union has called for government intervention to provide relief plans to mitigate the impact of fuel subsidy removal, which significantly affects their operational costs. They also demand a regulatory framework specific to the ride-hailing industry and access to special grants or loans for car financing to support their work.

9.5.2 *Fintech Companies and Ride-Hailing Drivers*

Union activities in Nigeria have primarily centered on addressing issues with the government and ride-hailing platforms related to commission fees, fuel costs, and regulatory protections. However, there has been a notable lack of discussion concerning the potential for exploitation and forced labor that could arise between Fintech companies such as Moove and ride-hailing drivers. Unlike the typically informal employment relationships found between drivers and ride-hailing platforms

Table 9.1 Select cases of workers' protests

Year	Month	Place	Leading organization	Purpose	Main causation	Scale
2024	Feb	Nigeria	Nigerian Association of Road Transport Owners (NARTO)	Appeal for subsidy	Rising operational costs due to devaluation of the domestic currency	Nationwide
2023	Feb	Lagos, Nigeria	Professional E-hailing Drivers and Private Owners Association (PEDPOA)	Reduces driver remittance to N5000/day	Forced Working hour and high mortgage payment	State-wide
2023	June	Nigeria	Amalgamated Union of App-based Transport Workers of Nigeria (AUATWON)	Lower the Commission Fee to 10%	High Platform Commission Fee, low pay	Nationwide
2022	June	Nigeria	AUATWON	Demanding paid overtime and holiday pay, pensions, social security, and union recognition	Rising operational costs due to subsidy cancellation	Nationwide
2021	April	Lagos, Nigeria	PEDPA	Increase the income	Low pay/ Low price of service	State-wide
2020	August	Lagos, Nigeria	National Union of Professional App-based Transport Workers and PEDPA	Lower the proposed service tax and the platform commission fee	Guidelines for Online Hailing Business Operation of Taxi	State-wide

such as Uber, the relationship with Moove involves a complex contractual agreement that defines their interaction as seller-buyer, lender-borrower, and maintenance service provider-customer. Under this contract, drivers acquire vehicles from Moove, often without a down payment, mainly for use in ride-hailing and delivery services. The amortization scheme, which can extend from 12 to 48 months, includes a specific interest rate that varies between 8% and 13%, depending on the vehicle model, amortization period, and other factors. To secure the mortgage, Moove requires drivers to make daily installment payments and mandates a minimum of 10 working hours per day, six days a week. According to a report published by Cambridge University, most of the drivers (78%) work for 12 h a day [25]. The exhaustion of long-time working further rises driver's concerns in fatigue driving and safety of working. While clauses about working hours are designed to protect Moove from payment defaults, they also complicate the relationship between Moove and the drivers, potentially blurring the lines between a lender-borrower dynamic and an underlying employer-employee relationship.

From a legal perspective, the International Labor Organization (ILO) has noted that the existence of an employment relationship should be determined based on the substantive nature of the relationship, particularly factors of subordination and dependence, rather than merely the presence or absence of a formal written contract [28]. While this determination can be complex in diverse practical scenarios, legal professionals often evaluate certain key characteristics to identify a substantial employment relationship. These characteristics typically include the level of supervision, disciplinary authority, economic dependency, the fixed duration of the engagement, and the exclusivity of the services provided.

In the case of Moove and its loan-bounding drivers, the stipulations concerning "working hours" and "penalties for default" introduce certain ambiguity to the relationships with drivers. As drivers' revenue is solely dependent on their driving activities, they are required to commit to full-time hours. These factors collectively fulfill the criteria indicative of an employment relationship, suggesting a level of subordination and economic dependence consistent with employer-employee dynamics. Despite the absence of a traditional written employment contract, these conditions suggest that Moove should assume the responsibilities and duties typical of an employer, providing the protections and obligations such a relationship entails [10].

The ILO defines "forced labor" based on two primary elements: the menace of penalty and involuntariness. The menace of penalty involves threats or actual penalties imposed to coerce someone to work against their will. Involuntariness refers to situations where individuals are forced, deceived, or coerced into working, and their freedom to choose employment or leave is significantly restricted or denied. The lack of free and informed consent by the worker exacerbates this issue. Such labor conditions place workers at a disadvantage, leaving them vulnerable to exploitation by profit-seeking employers or contractors. When comparing the conditions of ride-hailing drivers with the ILO definitions, there are clear overlaps in the use of coercion, penalties, and deception to enforce work, regardless of the economic and policy challenges that workers face.

For drivers whose earnings are solely dependent on platform-generated income, daily revenue can vary significantly. To enhance revenue stability and ensure full payment of installments, Moove requires drivers to meet specific Key Performance Indicators (KPIs); failure to do so may result in penalties or even vehicle seizure. Moreover, payments to Moove are automatically deducted from the drivers' accounts in their registered platform. However, the amount of payment may not be sufficiently transparent and well understood by the registered the drivers. On May 30, 2024, a small-scale protest erupted in Lagos, Nigeria, over a sudden increase in the daily repayment rate, escalating from 35% of gross earnings to ₦14,400 per day, which constitutes roughly 80% of the average daily income of ₦18,000 for drivers.

Apart from these unconsented payment schedule adjustments, Moove has also faced criticism for its ability to remotely lock vehicles and track driving routes. In May 2024, news emerged of an incident in Lagos where a driver experienced an abrupt vehicle lockout and seizure by Moove, triggered by a one-month default on payment obligations.

Lagos State has intensified its regulation of the ride-hailing sector through measures such as licensing requirements, service charge caps, vehicle standards, and mandatory data sharing. However, a recent news report highlights contentious interactions between the state and Moove [8]. According to the report, Lagos State officials have been impounding vehicles of Moove drivers due to the company's refusal to integrate its API with the state government's system. In a particularly aggressive enforcement tactic, Ministry of Transportation officials allegedly ordered rides only to deflate the tires and seize the vehicles upon arrival. This action came in response to a regulation that requires e-hailing companies to provide access to their databases, which Moove had declined to do. In response to these vehicle impoundments and the disruptions they caused, the AUATWON advised drivers to cease using the Moove app until the conflict is resolved.

9.6 Conclusion

This chapter delves into the disjunction between the widespread optimism driven by the rise of digital economy and the imperative to address critical issues of justice and inclusion amidst the burgeoning Fintech and platform economies in Africa. It scrutinizes the dynamics of partnerships between Fintech companies and ride-hailing platforms, providing insight into the complex web of stakeholders and relationships that influence urban economic transformations. While Fintech holds promise for addressing numerous long-standing challenges within the African context, a detailed examination of the day-to-day realities faced by ride-hailing drivers reveals how these technologies entrap cities and citizens in deeply inequitable and exploitative capitalist endeavors. This analysis underscores the need for a critical assessment of the contradictory impacts of digital technologies, urging continued vigilance to ensure Fintech serves as a force for equitable change, rather than entrenching existing power imbalances and wealth disparities.

Therefore, we strongly encourage local government to play a crucial role in facilitating multi-party dialogues among Moove, drivers, labor unions, platforms, and legislators, utilizing both formal and informal channels. The government should establish a robust mechanism to promote continuous and effective communications. Such dialogues are important to enhance mutual understanding of current challenges and problems. They are essential in forging a shared vision, especially during times of economic crisis. The goal of these discussions is to strengthen the overall resilience of Nigeria ride-hailing sector and to contribute to the development of optimized regulatory strategies for all stakeholders involved. Multi-party dialogues are particularly important for complex issues that affect multiple groups, including revenue allocation, data sharing, labor welfare, working conditions, budgeting and strategic planning of entities, and risk of compliance. The outcomes of these dialogues should be valued and leveraged to inform a rage of relevant policy changes. We also recommend the municipal authorities pay more close attention to the contractual relationship between Moove and drivers. A detailed examination of daily operational activities will allow municipal regulators to formulate better strategies for regulating the Fintech companies and lead to a more resilient and sustainable development of the ride-hailing sector.

Acknowledgements This research is partially funded by the 2024 Portman Family Graduate Award from Department of City and Regional Planning, Cornell. We acknowledge the generous support from the Portman Family.

References

1. Agbonifo J (2023) Fuel subsidy protests in Nigeria: the promise and mirage of empowerment. Extract Indus Soc 16
2. Alade I, Kavame Eroglu ZG (2023) Disruptive innovations or enhancing financial inclusion: what does Fintech mean for Africa? Vanderbilt J Trans Law 56(3). https://doi.org/10.2139/ssrn.4215020
3. Andreas Nienhaus SR, FO (2023) How shared mobility impacts the global urban landscape.
4. Bernards N (2022) Colonial financial infrastructures and Kenya's uneven Fintech boom. Antipode 54(3):708–728. https://doi.org/10.1111/anti.12810
5. Boateng FG, Appau S, Baako KT (2022) The rise of 'smart' solutions in Africa: a review of the socio-environmental cost of the transportation and employment benefits of ride-hailing technology in Ghana. Hum Social Sci Commun 9(1):1–11. https://doi.org/10.1057/s41599-022-01258-6
6. Cieslik K, Banya R, Vira B (2022) Offline contexts of online jobs: platform drivers, decent work, and informality in Lagos, Nigeria. Dev Policy Rev 40(4):1–19. https://doi.org/10.1111/dpr.12595
7. Cirolia LR, Sitas R, Pollio A, Sebarenzi AG, Guma PK (2023) Silicon Savannahs and motorcycle taxis: a Southern perspective on the frontiers of platform urbanism. Environ Plan A. https://doi.org/10.1177/0308518X231170193
8. Dosunmu D (2023) Uber's vehicle-financing partner is impounding drivers' cars in Nigeria. Rest World
9. Elas G Co (2020) Online hailing regulations in lagos state 2020: unconstitutional and uncertainty
10. Enwukwe NE (n.d.) The employment status of Nigerian workers in the gig economy: using Uber as a case study. J Policy Global 107(56)
11. Festival Godwin Boateng SA, KTB (2022) The rise of 'smart' solutions in Africa: a review of the socio-environmental cost of the transportation and employment benefits of ride-hailing technology in Ghana. Hum Social Sci Commun.
12. Gaschler F (2021) Fintech in Africa how digital payment tech is bringing financial services to the unbanked 44096:1–55
13. Kanife E (2024) Impounded vehicles: Uber says it is already sharing data with Lagos government, unsure about API integration. Technext
14. Kanife E (2023) Moove drivers in Nigeria are protesting against unbearable working conditions. Technext

15. Kene-Okafor T (2024) Uber leads $100M investment in African mobility Fintech Moove as valuation hits $750M. Techcrunch. https://afrotech.com/moove-uber-series-b-funding-round/
16. Kola-Oyeneyin T, Kuyoro M, Olanrewaju T (2020) Harnessing Nigeria's Fintech potential
17. Langley P, Leyshon A (2022) Neo-colonial credit: FinTech platforms in Africa. J Cult Econ 15(4):401–415. https://doi.org/10.1080/17530350.2022.2028652
18. Langley P, Rodima-Taylor D (2022) FinTech in Africa: an editorial introduction. J Cult Econ 15(4):387–400. https://doi.org/10.1080/17530350.2022.2092193
19. Mann L, Iazzolino G (2021) From development state to corporate Leviathan: historicizing the infrastructural performativity of digital platforms within Kenyan agriculture. Dev Chang 52(4):829–854. https://doi.org/10.1111/dech.12671
20. Ohuocha C (2024). Nigeria's inflation rises to more than 27-year high in December. Reuters
21. Okolie CN, Akwiwu I (2023) A review of the new Nigeria startup Act 2022: benefits and challenges for the tech ecosystem in Nigeria. Int J Social Sci Hum 13(5)
22. Oladunmade M (2023) Techcab "Moove drivers protest in Lagos over continued frustration", https://techcabal.com/2023/02/27/moove-drivers-protest-in-lagos/
23. Onyishi AO, Eme OI, Emeh IEJ (2012) Unemployment rate in Nigeria: agenda for government. Acad J Interdisc Stud 1(6). https://doi.org/10.5901/ajis.2014.v3n4p103
24. Pollio A (2019) Forefronts of the sharing economy: Uber in Cape Town. Int J Urban Reg Res 43(4):760–775. https://doi.org/10.1111/1468-2427.12788
25. Roland Banya KCODW (2022) Driving for platforms in Lagos and Abuja—can e-hailing platforms create decent jobs
26. Sitas R, Cirolia LR, Pollio A, Sebarenzi AG, Guma PK, Rajashekar A (2023) Platform politics and silicon Savannahs: Fintech and the platformed motorcycle: speculating on ordinary mobility economies in urban Africa (Issue June). https://doi.org/10.25375/uct.22820279
27. Statista (2023) Market insights—ride-hailing—Nigeria
28. Stefano V et al (2021) Platform work and the employment relationship ILO Working paper 27, ILO Working paper 27, International Labor Organization, ISSN: 2708–3446
29. The World Bank (2023) Nigeria development update—seizing the opportunity
30. What Has Happened To Nigeria's Economy Under Tinubu? Rewane Breaks Down The Numbers (2024) Channels television

Chapter 10
Diversity as Catalyst: A Future Built on Resilience Thinking

Ali Cheshmehzangi, Zaheer Allam, and Gaetan Siew

Abstract In this concluding chapter, we synthesize the key insights gathered from various case studies and discussions throughout the book. The chapter begins by framing the central view that diversity, when leveraged effectively, acts as a powerful driver for creating resilient and dynamic urban environments. It emphasizes that cities thrive not just by embracing diversity, but by integrating it into their strategic planning and development processes. The chapter starts with this general overview and then provides a reflection of all case study chapters included in the book. The chapter reflects on how each case study presented in the book illustrates the versatile ways in which diversity contributes to urban vitality, economic growth, and urban resilience. The contributions are diverse by nature and their focus of research but align very well with the United Nations' Sustainable Development Goals. The chapter reflectively highlights examples of how diverse cultural, ethnic, and socioeconomic backgrounds have enriched urban spaces, leading to increased creativity, economic growth, and social cohesion. The narrative ties together the findings from different cities, demonstrating that while challenges exist, the benefits of a diverse population are substantial and transformative.

Keywords Urban sustainability · Cultural diversity · Economic growth · Urban resilience · Sustainable development goals (SDG 11) · Urban planning · Inclusive cities

---

A. Cheshmehzangi (✉)
School of Architecture, Design and Planning (ADP), The University of Queensland, Brisbane, QLD, Australia
e-mail: A.Chesh@uq.edu.au

Z. Allam
Urban Transformations Research Centre (UTRC), Western Sydney University, Parramatta, NSW, Australia

School of Architecture, Design and Planning, The University of Queensland, Brisbane, QLD, Australia

G. Siew
International Union of Architects (UIA), Paris, France

Global Creative Leadership Initiative (GCLI), Port Louis, Mauritius

G. Siew et al. (eds.), *Diversity as Catalyst: Economic Growth and Urban Resilience in Global Cityscapes*, Urban Sustainability, https://doi.org/10.1007/978-981-97-8776-0_10

10.1 A Brief Overview

In an era of rapid urbanization, sustainability poses significant challenges such as resource management, environmental protection, and social equity [3, 4, 18]. Cultural diversity is highlighted as a vital element in addressing these challenges [10]. Diverse perspectives and practices enrich urban populations, enhancing creativity and problem-solving, which leads to more resilient urban systems [28]. The chapter underlines that cities with high cultural diversity enjoy a broader range of ideas, traditions, and approaches, fostering a dynamic culture that attracts talent, stimulates economic activity, and improves residents' quality of life.

Urban resilience is another key theme covered throughout the book. Diverse cities can better withstand and recover from various shocks such as economic downturns, natural disasters, and social conflicts [2, 5, 12, 17, 22, 27]. It is therefore arguable that diversity enhances resilience through social capital and community networks [9, 23, 26], which are essential for coping with crises [11, 13, 24]. As a result, inclusive and socially integrated cities are stronger and more sustainable in the long run [6, 16, 20]. The chapter aligns with the United Nations' Sustainable Development Goal 11, which aims to make cities inclusive, safe, resilient, and sustainable. It emphasizes the importance of integrating cultural diversity into urban planning and development processes. Effective urban planning should consider the diverse needs of urban populations, ensuring inclusive and equitable spaces [7]. Participatory planning processes that engage diverse communities [8, 14, 15] lead to more responsive and sustainable outcomes.

As an overview, the case study chapters of the book align with the United Nations' Sustainable Development Goal 11, which aims to make cities inclusive, safe, resilient, and sustainable. The central focus of the book is to emphasize the importance of integrating diversity into urban planning and development processes [1, 19, 21, 25]; hence, regarding diversity as a catalyst for cities and urban environments/communities. Effective urban planning should consider the diverse needs of urban populations, ensuring inclusive and equitable spaces. Therefore, a major overarching take out is related to participatory planning processes that engage diverse communities lead to more responsive and sustainable outcomes.

10.1.1 Diversity, Cities, and Meeting the SDG 11

Diversity is highly relevant to Sustainable Development Goal 11 (SDG 11), which aims to make cities and human settlements inclusive, safe, resilient, and sustainable (see Fig. 10.1). Here is how diversity plays a critical role in achieving this goal:

Fig. 10.1 Relevance of diversity to SDG 11

Inclusive Urban Development

Diverse Perspectives and Innovation: Diversity brings together people with various backgrounds, experiences, and perspectives, fostering innovation and creative solutions to urban challenges. Inclusive urban planning that considers diverse needs can lead to more effective and equitable outcomes.

Representation and Equity: Ensuring diverse representation in decision-making processes helps address the needs of all community members, especially marginalized groups. This can lead to policies and practices that promote equity and justice in urban settings.

Social Cohesion and Community Resilience

Social Cohesion: Diverse communities tend to have stronger social networks and a greater sense of community. This social cohesion is vital for resilience, as it helps communities to support each other in times of crisis and work together towards common goals.

Cultural Vibrancy: Diversity enriches the cultural life of cities, making them more vibrant and attractive places to live and visit. Cultural diversity can also contribute to the local economy through tourism and creative industries.

Economic Growth and Opportunities

Economic Dynamism: Diversity can drive economic growth by attracting a diverse workforce and fostering an environment where new ideas and businesses can thrive. This can lead to more job opportunities and economic resilience.

Inclusive Growth: By ensuring that all segments of the population can participate in and benefit from economic opportunities, cities can promote inclusive growth and reduce inequalities.

Sustainable Urban Environments

Environmental Justice: Diverse communities often advocate for environmental justice, ensuring that environmental benefits and burdens are shared equitably. This can lead to more sustainable urban environments that are healthier for all residents.

Adaptive Capacity: Diverse communities are often more adaptable to change, whether it's economic shifts, technological advancements, or climate impacts. This adaptability is crucial for building resilient and sustainable cities.

Health and Well-being

Public Health: Diversity in cities can lead to more comprehensive public health strategies that address the specific needs of various populations, improving overall health outcomes.

Quality of Life: A diverse urban environment can enhance the quality of life by providing a wide range of services, amenities, and cultural experiences that cater to different preferences and needs.

10.2 Reflection on Case Study Chapters

Here are some key points and take outs from case study chapters of the book:

Chapter 2: Designing the institutional framework of Hybrid Special Economic Zones (HEZ)

Economic resilience in cities and countries is crucial for maintaining stability and growth during economic shocks. Special economic zones (SEZs) have become significant in economic growth strategies, particularly through large-scale industrial projects in oil, gas, and services. However, these zones often focus on either export or domestic markets, lacking a diversified approach. This narrow focus can undermine a city's resilience. To address this, hybrid special economic zones, which integrate both export-oriented and market-oriented activities, are proposed. These zones emphasize diversity in their economic structures to bolster resilience across economic, social, cultural, and environmental dimensions. Bushehr province in Iran, home to major oil-related industries, serves as a case study. The region's economy, heavily reliant on exports, is vulnerable to international market fluctuations and domestic policies. The chapter suggested a hybrid model for SEZs, incorporating sustainability and growth-oriented policies to enhance resilience. This model aims to internalize the economy while pursuing external market opportunities, ultimately strengthening economic stability.

Lessons Learnt

1. **Diversification Enhances Resilience**: A diverse economic structure is crucial for cities to withstand economic shocks. Relying on a single industry or market can make cities vulnerable to fluctuations.
2. **Hybrid Economic Zones**: Combining export-oriented and market-oriented activities within special economic zones can provide a balanced and resilient economic foundation.
3. **Institutional Framework**: Effective policies and institutional frameworks are necessary to support the dual focus on sustainability and growth in hybrid economic zones.
4. **Case Study Insight**: The economic situation of Bushehr province highlights the risks of dependency on a single sector and the importance of diversification for economic stability.
5. **Policy Recommendations**: The chapter emphasizes the need for comprehensive strategies that integrate internal and external economic approaches to enhance the overall resilience of cities and countries.

Chapter 3: Temporary Appropriation as a Catalyst for Diversity in Urban Development

This chapter investigated the concept of temporary appropriation (TA) and its significant role in enhancing urban diversity. Temporary appropriation refers to the practice of individuals or communities repurposing urban spaces for alternative uses, reflecting the dynamic interaction between social groups and the built environment.

By examining case studies from cities like Venice, Mexico City, Hong Kong, and Auckland, the research highlights how TA fosters inclusivity, stimulates economic activities, and strengthens community resilience. It highlights the importance of TA in integrating cultural expressions, community gatherings, and grassroots economic initiatives into urban spaces. The chapter argued that TA is a crucial tool for urban planners and policymakers, promoting flexible urban design strategies that meet the evolving needs and creative expressions of diverse populations. This approach encourages a more inclusive, sustainable, and resilient urban future.

Lessons Learnt

1. **Enhancing Social Cohesion**: Temporary appropriation promotes social cohesion by enabling diverse social groups to interact and engage with urban spaces creatively and inclusively.
2. **Economic Stimulus**: Repurposing urban spaces through TA can stimulate local economies by supporting grassroots economic initiatives and cultural activities.
3. **Community Resilience**: TA strengthens community resilience by fostering a sense of ownership and engagement among residents, contributing to more vibrant and adaptive urban environments.
4. **Flexible Urban Design**: Urban planners and policymakers should adopt flexible design strategies that accommodate the dynamic and evolving needs of urban populations, recognizing the value of TA.
5. **Global Relevance**: The positive impacts of TA observed in diverse cities worldwide highlight its universal applicability and potential to enhance urban diversity and resilience globally.

Chapter 4: Assessing the Position of the Neighbourhood-oriented Management Approach in the Economic Management of Medium-sized Cities

This chapter emphasized the importance of neighbourhood-oriented management in the economic governance of medium-sized cities. Neighbourhoods are vital to urban identity and effective city management. However, contemporary urban management often overlooks citizen engagement, impacting economic management negatively. The research focused on assessing managers' attention to neighbourhood potentials in economic management in Bahar and evaluating the strengths and weaknesses of this approach. Data was gathered using a descriptive-analytical method and questionnaires, involving 196 urban management experts in Bahar.

The findings indicate that current urban management practices are suboptimal, particularly in addressing complex economic issues. Thus, a neighbourhood-oriented approach is crucial for economic development and optimal city management, serving as a bridge between local communities and central governance. A single-sample t-test highlighted economic transparency as a key urban management component, with significant efforts by Bahar Municipality to inform the public about its activities. The central neighbourhood of Bahar was found to have an average and appropriate level of economic management.

Lessons Learnt

1. **Neighbourhood Importance:** Effective urban management should prioritize neighbourhoods as fundamental units, fostering a strong sense of identity and engagement among residents.
2. **Economic Transparency:** Clear communication of municipal activities enhances trust and governance efficiency.
3. **Neighbourhood-Oriented Approach:** Integrating local perspectives into economic management can address complex urban issues more effectively.
4. **Citizen Engagement:** Active involvement of citizens in urban management is crucial for sustainable economic growth and development.

Chapter 5: 15-minute Cities: A Systems Thinking Analysis of a New Urban Development Framework

The 15-minute City is a modern urban model that gained prominence during the COVID-19 pandemic, focusing on accessibility, sustainability, and health. It envisions a city where all essential services, including groceries, medical facilities, childcare, education, and employment, are within a 15-minute walk or bike ride from any household. This model promotes the development of self-sufficient neighbourhood units within larger cities. Although the concept of a well-serviced, dense urban lifestyle is not new, the 15-minute City framework provides a fresh perspective to improve city planning and enhance the quality of life.

The chapter utilized a Systems Thinking approach, employing tools like DSRP (Distinctions, Systems, Relationships, Perspectives), POSIWID (Purpose of the System is What it Does), and CAS (Complex Adaptive System) analyses to evaluate the effectiveness, opportunities, and limitations of the 15-minute City model. The study includes case studies of Dubai's proposed 20-minute City and Barcelona's implementation of the 15-minute City, offering insights into the practical application of the framework across different socio-cultural, economic, and political contexts. This comprehensive analysis aims to contribute to the discourse on sustainable urban development by exploring the feasibility and adaptability of the 15-minute City model.

Lessons Learnt

1. **Proximity Principle:** The 15-min City model emphasizes the importance of proximity in urban planning, enhancing accessibility to essential services and improving quality of life.
2. **Systems Thinking:** Utilizing a Systems Thinking approach helps identify potential blind spots and improve the overall framework for real-world implementation.
3. **Global Adaptability:** The model's principles can be adapted to various socio-cultural, economic, and political contexts, as demonstrated by the case studies of Dubai and Barcelona.
4. **Sustainability Focus:** Prioritizing local access to services supports sustainable urban development by reducing the need for long commutes and lowering carbon emissions.

5. **Holistic Planning:** Integrating diverse analytical tools and case studies provides a comprehensive understanding of the model's strengths and challenges, guiding future urban planning efforts.

Chapter 6: Urban horizons in China: challenges and opportunities for community intervention in a country marked by the Heihe-Tengchong Line

This chapter delved into urbanization and population density trends in China from 2000 to 2020, focusing on three primary objectives. Firstly, it investigated how state policies have shaped the real estate market and urban sprawl. Secondly, it examined changes in population density using annual census data to understand the impact of these policies on demographic distribution. Thirdly, it explored the relationship between national policies and population density in the context of community policies. The study utilized data from over 5000 urban areas to assess how policies influence spatial population distribution.

The chapter highlighted the historical and ongoing impact of Chinese community policies, such as efforts for balanced regional development, hukou system reforms to enhance mobility and rural–urban integration, and strategies for sustainable urbanization. These policies aim to improve living standards and urban infrastructure, tackling challenges related to urban concentration while fostering inclusive and balanced development. However, the long-term outcomes of these policies remain uncertain.

Lessons Learnt

1. **Policy Impact on Urbanization:** State policies significantly influence real estate markets and urban sprawl, highlighting the importance of strategic planning in managing urban growth.
2. **Population Density Distribution:** Annual census data reveals how policies affect demographic shifts, emphasizing the need for policies that promote balanced population distribution.
3. **Community Policy Integration:** Integrating national policies with community-level initiatives is crucial for achieving sustainable urbanization and improving living standards.
4. **Balanced Regional Development:** Efforts to balance regional development through policy reforms can enhance rural–urban integration and mobility, although their effectiveness varies.
5. **Future Policy Recommendations:** Continuous assessment and adaptation of policies are necessary to address emerging urbanization challenges and to foster inclusive growth and sustainable development.

Chapter 7: Rereading the Role of Cafés as Urban Catalysts with a Focus on the Case of Tehran, Iran

This chapter explored the evolving role of cafés as catalysts for urban and social change, tracing their history from traditional coffee houses in 16th-century Iran to modern-day establishments. Historically, these spaces have served diverse functions, from hosting royalty and lower-class gatherings to becoming centers for artists and intellectuals. Despite facing governmental scrutiny and restrictions, cafés have persisted, reflecting societal shifts and resilience. The chapter employs a case study method, utilizing historical texts, travelogues, maps, and spatial data to analyze the transformation of cafés in Tehran. It highlights how these hybrid spaces have continually adapted to social changes, from the Safavid era through the Qajar and Pahlavi periods, to the present day. Recent political pressures have led to the closure of many cafés, yet new ones continue to emerge, contributing to urban regeneration and social diversity.

Lessons Learnt

1. **Cafés as Social Hubs:** Throughout history, cafés have served as crucial social gathering spots, fostering connections across different social classes and cultural backgrounds.
2. **Resilience and Adaptation:** Despite political and social challenges, cafés have shown remarkable resilience, continually reopening and evolving to meet the needs of society.
3. **Cultural Significance:** Cafés have played a significant role in cultural exchange and intellectual discourse, especially during periods of political and social upheaval.
4. **Urban Regeneration:** The rise of new café styles contributes to urban renewal, revitalizing neglected areas and transforming them into vibrant socio-commercial spaces.
5. **Centres of Diversity and Freedom:** Modern cafés act as centers of diversity and freedom, promoting social cohesion and providing a platform for expressing diverse lifestyles.
6. **Economic Impact:** Beyond their social function, cafés represent a robust business market, driving economic activity and supporting urban economies.

Chapter 8: Brownfields in China: the Case of Liaocheng Chemical Plant

In recent decades, the approach to brownfield regeneration has become crucial for managing urban sprawl and efficiently using natural resources. This policy is well-established in many developed countries but is still developing in nations like China. The Chinese government is working to establish a nationwide framework for brownfield redevelopment through pilot projects, despite the absence of a comprehensive legislative structure. This study examined the regeneration of a specific industrial brownfield in eastern China, the Liaocheng Chemical Plant, which has been heavily utilized for decades. The research highlights China's efforts to phase out outdated and polluting industries while implementing environmental remediation to rehabilitate contaminated land. By analyzing this case, including personal interviews and site visits, the study provided insights into China's evolving policies and regulatory efforts to address and manage land contamination within the broader context of regional and political changes.

Lessons Learnt

1. **Importance of Brownfield Regeneration:** Effective brownfield regeneration is crucial for managing urban sprawl and promoting sustainable land use.

2. **Legislative Challenges:** Many countries, including China, face challenges in developing comprehensive regulations for brownfield redevelopment.
3. **Pilot Projects as Testing Grounds:** Pilot studies can play a critical role in shaping and refining national frameworks for brownfield management.
4. **Environmental Remediation:** Addressing contamination through environmental remediation is essential for rehabilitating brownfield sites.
5. **Contextual Analysis:** Understanding brownfield regeneration requires considering the broader regional and political context to grasp its impact fully.
6. **Practical Application of Policies:** Real-world case studies are vital for evaluating the effectiveness of evolving policies and regulatory measures in managing land contamination.

Chapter 9: Fintech as Developmental Catalyst or Exploitative Mechanism? A Case Study of Moove in Lagos, Nigeria

This chapter explored the impact of fintech on urban economies and living experiences in Africa, specifically through the lens of the ride-hailing sector in Nigeria. It examined how fintech innovations, exemplified by the collaboration between Moove, a fintech company, and Uber, a leading ride-hailing platform, are reshaping urban mobility in Lagos. By providing new vehicle financing solutions, fintech helps more Nigerians engage in the digital economy and access ride-hailing opportunities.

However, the chapter also highlighted the downside of this collaboration. While fintech facilitates access to vehicle ownership, it also intensifies job insecurity among drivers, particularly during economic downturns or policy shifts. This dual impact emphasizes the complexities of fintech-platform partnerships, which, while expanding financial inclusion and urban mobility, also introduce new vulnerabilities for workers.

Lessons Learnt

1. **Fintech's Dual Role:** Fintech can drive economic inclusion and access to services, but it can also create new forms of economic instability.
2. **Policy Considerations:** There is a need for policies that balance the benefits of fintech with protections for vulnerable workers.
3. **Sector-Specific Impacts:** The effects of fintech collaborations are nuanced and vary across sectors, requiring tailored solutions to address both opportunities and challenges.

10.3 A Summary

This chapter serves as a concluding chapter to the book focused on understanding the vital interplay between diversity and cities, by predominantly considering "diversity as catalyst", which offers strategic guidance for fostering vibrant and inclusive cities.

Through case study examples, the book delves deeply into several key themes related to the integration of diversity in urban development. Firstly, it explores how diversity fuels innovation by bringing together varied perspectives, which can lead to creative problem-solving and novel solutions. This diversity of thought enhances the ability of cities to adapt to new challenges and opportunities, driving technological advancement and cultural enrichment. Secondly, the book highlights how a diverse population can stimulate economic activity; and essentially, economic growth. Diverse communities often lead to a more vibrant local economy by attracting businesses, promoting entrepreneurial ventures, and creating a dynamic marketplace that reflects a wide array of needs and preferences. This economic boost contributes to overall urban prosperity and competitiveness.

Additionally, the book addresses how diversity strengthens resilience, and particularly, social resilience. Cities with diverse populations tend to have more robust social networks and support systems, which can help communities better withstand and recover from economic downturns, natural disasters, and social conflicts. Overall, this collection highlights the importance of inclusivity in urban planning, asserting that effectively managing diversity is not merely about addressing immediate challenges but about fostering long-term sustainability and equity. It emphasizes that inclusive urban planning practices can lead to more balanced and fair development outcomes, ensuring that all community members benefit from urban growth.

In the final section, we present strategic recommendations for policymakers, urban planners, and community leaders. We emphasize the need for policies that actively foster inclusivity and leverage the strengths of diverse populations to create cities that are both more resilient and more vibrant. The chapter ends with 'a call to action', urging stakeholders to view diversity as a catalyst for positive transformation and urban progress, rather than a challenge to be addressed. Embracing this approach will allow cities to fully tap into the potential of their diverse communities and pave the way for a more equitable and dynamic urban future.

References

1. Adams D, Tiesdell S, White JT (2013) Smart parcelization and place diversity: reconciling real estate and urban design priorities. J Urban Des 18(4):459–477
2. Albrechts L (2005) Creativity as a drive for change. Plan Theory 4(3):247–269
3. Allam Z, Cheshmehzangi A (2024) The green new deal as a pathway for climate and social justice. In: Sustainable futures and green new deals. Springer Nature Switzerland, Cham, pp 105–133
4. Allam Z, Cheshmehzangi A, Jones DS (2023) Climate and social justice in the political landscape of urban resilience and mercantilism: a concluding note. In: Climate and social justice: the political economy of urban resilience and mercantilism. Springer Nature Singapore, Singapore, pp 103–110
5. Arroyo KK (2017) Creative policymaking: taking the lessons of creative placemaking to scale. Artivate 6(2):58–72
6. Belsky ES, DuBroff N, McCue D, Harris C, McCartney S, Molinsky J (2013) Advancing inclusive and sustainable urban development: correcting planning failures and connecting communities to capital. Joint Center for Housing Studies of Harvard University, Cambridge, MA, USA
7. Burayidi MA (ed) (2015) Cities and the politics of difference: multiculturalism and diversity in urban planning. University of Toronto Press, Toronto

8. Chen W, Cheshmehzangi A, Yu J, Mangi E, Heath T, Zhang Q (2023) An analysis of patterns of public engagement in China's community micro-rehabilitation projects: a case study of Guangzhou. World Dev Sustain 3:100108
9. Cheshmehzangi A (2014) Spatial syntagma and identity of a place: sensing, relating to, and knowing a place. J Hum Behav Soc Environ 24(7):799–810
10. Cheshmehzangi A (2020) Identity of cities and city of identities. Springer Nature, Singapore
11. Cheshmehzangi A (2023) Urban regeneration and building colours: mapping visual and perceptual dimensions of Dali's old towns. In: Mapping urban regeneration: city life experiences in Yunnan, China. Springer Nature Singapore, Singapore, pp 105–137
12. Cheshmehzangi A, Dawodu A (2019) Introduction: climate change and cities—perspectives, planning and people. In: Sustainable urban development in the age of climate change: people: the cure or curse, pp 1–22
13. Cheshmehzangi A, Sedrez M, Dawodu A, Heath T, Zhao H, Li T (2023) Urban resilience to future pandemics. In: Resilience vs pandemics: innovations in cities and neighbourhoods. Springer Nature Singapore, Singapore, pp 1–12
14. Cheshmehzangi A, Zou T, Tang T, Xie L (2023) Exploring influential factors for decision-making in urban development: urban heritage protection and displacing living communities
15. Dawodu A, Sharifi A, Cheshmehzangi A, Oladejo J (2021) The illusion of participation: are participatory indicators truly effective in neigborhood sustainability assessment tools. J Clean Prod 311:127538
16. Elias P (2020) Inclusive city, perspectives, challenges, and pathways. Sustainable cities and communities. Springer International Publishing, Cham, pp 290–300
17. Fabbricatti K, Boissenin L, Citoni M (2020) Heritage community resilience: towards new approaches for urban resilience and sustainability. City Territory Arch 7(1):17
18. Flint RW, Danner MJ (2001) The nexus of sustainability & social equity: Virginia's Eastern Shore (USA) as a local example of global issues. Int J Econ Dev 3(2)
19. Greenlee AJ, Jackson A, Garcia-Zambrana I, Lee CA, Chrisinger B (2022) Where are we going? Where have we been? The climate for diversity within urban planning educational programs. J Plan Educ Res 42(3):331–349
20. Kohon J (2018) Social inclusion in the sustainable neighborhood? Idealism of urban social sustainability theory complicated by realities of community planning practice. City Cult Soc 15:14–22
21. Lu L, Jia B (1998) Continuity, integration and diversity: sustainable urban design concepts for redeveloping. Asian Geogr 17(1–2):41–54
22. Moglia M, Cork SJ, Boschetti F, Cook S, Bohensky E, Muster T, Page D (2018) Urban transformation stories for the 21st century: insights from strategic conversations. Glob Environ Chang 50:222–237
23. Smith JW, Anderson DH, Moore RL (2012) Social capital, place meanings, and perceived resilience to climate change. Rural Sociol 77(3):380–407
24. Su Z, Zhang H, McDonnell D, Ahmad J, Cheshmehzangi A, Yuan C (2022) Crisis communication strategies for health officials. Front Public Health 10:796572
25. Underwood DA, Hackney DD, Friesner D (2015) Criteria for sustainable community economic development: integrating diversity and solidarity into the planning process. J Econ Issues 49(4):1112–1123
26. Western J, Stimson R, Baum S, Van Gellecum Y (2005) Measuring community strength and social capital. Reg Stud 39(8):1095–1109
27. Ziafati Bafarasat A, Cheshmehzangi A, Ankowska A (2023) A set of 99 healthy city indicators for application in urban planning and design. Sustain Dev 31(3):1978–1989
28. Zellner M, Campbell SD (2015) Planning for deep-rooted problems: what can we learn from aligning complex systems and wicked problems? Plan Theory Pract 16(4):457–478

The manufacturer's authorised representative in the EU is Springer Nature Customer Service Centre GmbH, Europaplatz 3, 69115 Heidelberg, Germany. If you have any concerns regarding our products, please contact ProductSafety@springernature.com

Printed and bound by CPI Group (UK) Ltd, Croydon, CR0 4YY
01/10/2026
02225858-0003